XI All-Russian Scientific and Practical Conference "Automation Systems in Education, Science and Production 2017"

AA006284

IOP Conference Series: Materials Science and Engineering
Volume 354

Novokuznetsk, Russia
14 – 16 December 2017

ISBN: 978-1-5108-6284-5
ISSN: 1757-8981

Printed from e-media with permission by:

Curran Associates, Inc.
57 Morehouse Lane
Red Hook, NY 12571

CURRAN ASSOCIATES INC. proceedings.com

Some format issues inherent in the e-media version may also appear in this print version.

Copyright© (2017) by the Institute of Physics
All rights reserved. The material featured in this book is subject to IOP copyright protection, unless otherwise indicated.

Printed by Curran Associates, Inc. (2018)

For permission requests, please contact the Institute of Physics at the address below.

Institute of Physics
Dirac House, Temple Back
Bristol BS1 6BE UK

Phone: 44 1 17 929 7481
Fax: 44 1 17 920 0979

techtracking@iop.org

Additional copies of this publication are available from:

Curran Associates, Inc.
57 Morehouse Lane
Red Hook, NY 12571 USA
Phone: 845-758-0400
Fax: 845-758-2633
Email: curran@proceedings.com
Web: www.proceedings.com

Table of contents

Volume 354

XI All-Russian Scientific and Practical Conference (with international participation) "Automation systems in education, science and production, 2017"

14–16 December 2017, Novokuznetsk, Russian Federation

Accepted papers received: 20 April 2018
Published online: 10 May 2018

Preface

XI All-Russian Scientific and Practical Conference (with international participation) "Automation systems in education, science and production, 2017"

Peer review statement

Papers

About development of automation control systems

L P Myshlyaev, K G Wenger, K A Ivushkin and V N Makarov.....1

Precedent approach to the formation of programs for cyclic objects control

S M Kulakov, V B Trofimov, A S Dobrynin and E N Taraborina.....7

Incentive schemes in development of socio-economic systems

V V Grachev, K A Ivushkin and L P Myshlyaev.....16

Investigation of multidimensional control systems in the state space and wavelet medium

D B Fedosenkov, A A Simikova and B A Fedosenkov.....27

Development of similarity theory for control systems

L P Myshlyaev, V F Evtushenko, K A Ivushkin and G V Makarov.....33

Improving reliability of aggregation, numerical simulation and analysis of complex systems by empirical data

Boris S Dobronets and Olga A Popova.....38

About the algorithm for construction of coordinated university timetables

A S Dobrynin, S M Kulakov and E N Taraborina.....45

Problems in modernization of automation systems at coal preparation plants

L P Myshlyaev, M V Lyakhovets, K G Venger, I A Leontiev, G V Makarov and A S Salamatin.....49

Modernization of the automation control system of technological processes at the preparation plant in the conditions of technical re-equipment

M V Lyakhovets, K G Wenger, L P Myshlyaev, M V Shipunov, V V Grachev, M Yu Melkozerov and Sh A Fairoshin.....52

Experience of creating a multifunctional safety system at the coal mining enterprise

V V Reshetnikov, K S Davkaev, M V Korolkov and M V Lyakhovets.....58

About the mechanism of ERP-system pilot test

V V Mitkov and V V Zimin.....64

Analysis of the environmental and economic indicators of the industrial enterprise

V G Mikhailov and T V Kiseleva.....70

Management mechanisms for development of personnel professional competencies at a loading and haulage company

S V Shishkina, Yu D Pristupa, L D Pavlova and V N Fryanov.....76

Technical devices of powered roof support for the top coal caving as automation objects

M S Nikitenko, S A Kizilov, P I Nikolaev and I S Kuznetsov.....82

Longwall top coal caving (LTCC) mining technologies with roof softening by hydraulic fracturing method

V Klishin, S Nikitenko and G Opruk.....86

Concept of mobile operator position based on neurocomputer interface and augmented reality

S A Kizilov, M S Nikitenko and B Neogi.....91

Equivalent model of a dually-fed machine for electric drive control systems

I Yu Ostrovlyanchik and I Yu Popolzin.....95

Development of the pump protection system against cavitation on the basis of the stator current signature analysis of drive electric motor

M V Kipervasser, A V Gerasimuk and V P Simakov.....101

Features of the use of time-frequency distributions for controlling the mixture-producing aggregate

D B Fedosenkov, A A Simikova and B A Fedosenkov.....107

IOP Conf. Series: Materials Science and Engineering **354** (2018) 011001 doi:10.1088/1757-899X/354/1/011001

Preface

Since 1995 Siberian State Industrial University (SibSIU) in cooperation with the research and design institutes, educational institutions and industrial enterprises, social and economic structures has been organizing the All-Russian Scientific and Practical Conference (with international participation) "Automation systems in education, science and production".

The conference is held every two years and covers a wide range of issues related to the creation of automation systems for industrial and socio-economic complexes. The program and organizational committee of the conference includes well-known Russian scientists and developers of automation systems, as well as representatives of Belarus and Kazakhstan. The financial support of the conference is provided by the Russian Foundation for Basic Research, large industrial associations such as Consolidated Company "Sibshakhtostroy" LLC (Novokuznetsk), "Stroyservis" CJSC (Kemerovo).

In 2017 the conference was held on December 14-16. More than 100 papers were submitted by the scientists and specialists from Moscow, Voronezh, Stary Oskol, Yekaterinburg, Omsk, Novosibirsk, Kemerovo, Krasnoyarsk, Prokopyevsk, Novokuznetsk, representing academic, research and development, educational and design institutions, administrative and socio-economic structures, enterprises of various profiles.

The work of the conference was organized according to sections: the plenary session, actual problems of theory and practice of automation, automation and computer systems for production, educational and research purposes. A round table was held on the problems of automation in coal industry. Students, postgraduate students and young scientists took an active part in the conference.

The Organizing Committee thanks the Russian Foundation for Basic Research (project No. 17-07-20581), enterprises and organizations for financial support, as well as all conference participants for their contribution to the development of theory and practice of automation control systems.

Chairman of the Conference Organizing Committee
Doctor of Technical Sciences, Professor,
Rector of SibSIU E.V. Protopopov

Content from this work may be used under the terms of the Creative Commons Attribution 3.0 licence. Any further distribution of this work must maintain attribution to the author(s) and the title of the work, journal citation and DOI.
Published under licence by IOP Publishing Ltd

Peer review statement

All papers published in this volume of *IOP Conference Series: Materials Science and Engineering* have been peer reviewed through processes administered by the proceedings Editors. Reviews were conducted by expert referees to the professional and scientific standards expected of a proceedings journal published by IOP Publishing.

Content from this work may be used under the terms of the Creative Commons Attribution 3.0 licence. Any further distribution of this work must maintain attribution to the author(s) and the title of the work, journal citation and DOI.
Published under licence by IOP Publishing Ltd

IOP Conf. Series: Materials Science and Engineering **354** (2018) 012001 doi:10.1088/1757-899X/354/1/012001

About development of automation control systems

L P Myshlyaev[1], K G Wenger[2], K A Ivushkin[3] and V N Makarov[2]

[1]Siberian State Industrial University, 42 Kirova street, Novokuznetsk, 654007, Russia
[2]"Stroyservis" CJSC, 121 Kuznetsky ave., Kemerovo, 650055, Russia
[3]Consolidated Company "Sibshakhtostroy" LLC, 9 Kuznetskoye highway, Novokuznetsk, 654034, Russia

E-mail: mail@nicsu.ru

Abstract. The shortcomings of approaches to the development of modern control automation systems and ways of their improvement are given: the correct formation of objects for study and optimization; a joint synthesis of control objects and control systems, an increase in the structural diversity of the elements of control systems. Diagrams of control systems with purposefully variable structure of their elements are presented. Structures of control algorithms for an object with a purposefully variable structure are given.

1. Introduction

The increase in the indicators efficiency of industrial and social systems is impossible without qualitatively new approaches and methods of automation control. To date, a sufficiently high level of technical means of automation and basic (service) software has been achieved. The development of new directions in automation of control and methods and algorithms acceptable for practical use did not meet the requirements. There are several reasons for this situation and ways to improve automation control systems.

The problem setting and identification of the complete object under study and optimization are the first and most important steps in the problem solving under uncertainty. The greatest difficulty is the correct selection (formation) of a complete object of study and optimization. At this stage, the greatest number of errors is allowed, which, of course, leads to erroneous, including qualitatively, solutions. General principles for the identification of a complete object can serve the principles of system analysis, formulated in [1], where it is stated that the complete object should not be covered by strong external feedbacks. All feedbacks should be included in the object under study.

The examples of incorrect selection of a complete object are:

1) identification of objects in control systems based on the data of normal operation [2];

2) prediction of states and outputs of control objects when using the prediction results for generating control actions [3].

From these positions a revision of approaches to the creation of industrial complexes and training of specialists – engineers are needed. Traditionally, when creating industrial and often social complexes, technological diagrams (processes, units) are first developed, and only then control subsystems. This violates the principles of system analysis and, as a result, the optimal performance of the complex is not achieved. Achievement of the given dynamic properties of the complex is assigned to the control part of the system. Often this is either impossible due to the structure and parameters of

Content from this work may be used under the terms of the Creative Commons Attribution 3.0 licence. Any further distribution of this work must maintain attribution to the author(s) and the title of the work, journal citation and DOI.
Published under licence by IOP Publishing Ltd

IOP Conf. Series: Materials Science and Engineering **354** (2018) 012001 doi:10.1088/1757-899X/354/1/012001

the control object, or significantly complicates the control subsystem, especially its intellectual component.

2. Directions and approaches to the problem solving

More rational is the path of joint synthesis of the control object and the control subsystem. This requires a rethinking of methods and approaches to design of industrial complexes, development of appropriate methodologies and regulations, and it is also necessary to introduce changes in training and professional development programs for specialists, where the focus would be on the automated production (social) complexes, rather than on production processes (technologies) and control automation subsystems.

One of the promising options for creating efficient industrial complexes can lie in the way of increasing the structural diversity of control objects and control systems.

The developed theory of ***control systems with a variable structure*** presupposes changes in the structure of control system only, depending on the state of the control object and/or external conditions [4, 5]. Theoretical studies with a purposeful change in the structure of control object are only at the initial stage. In practice such systems are encountered very often. Here the following examples can be given: rolling production of a metallurgical enterprise, when the metal rolling route is first formed, that is, the structure of the control object is formed, and then the technological modes of rolling are realized; coal enrichment, when, depending on the properties of coal, technological chains and corresponding technological units are first determined, and then processes of coal enrichment with the possible operative replacement of the initial structure elements are carried out within this structure. Such systems can also include systems of competitive selections (tenders) for performance of research, design and experimental, construction and installation works on creation of industrial and social complexes. From the participants the performers are selected (the structure of the object is formed), the number of which can vary during the course of the project.

3. The generalized diagram of the object control system with a purposeful variable structure

The generalized diagram of the object control system with a purposeful variable structure can be represented in figure 1 [6]. Three types of implementation of this diagram are possible.

1. With the introduction of additional object structures to expand the range of permissible states and the choice of a sequence of structures to meet the specified requirements for control effectiveness.

2. With the current evaluation and analysis of performance indicators of control, the choice of the necessary structure of control object and provision of smooth control modes.

3. With the identification of type representative situations of the control system functioning for each system structure, imitating recalculation, estimation of control effectiveness indicators, selection of the necessary structure of the control object and provision of smooth control modes.

With such implementation of diagrams at the design stage it is necessary to include the option of changing the structure of the object.

4. The algorithm for object control with a purposefully variable structure

The most universal and effective for practice is the third option of the diagram implementation from figure 1 with the identification of type representative situations (TRS) and simulation of all structures of control system. The algorithm for implementation of such diagram is shown in figure 2 [6].

The absence of adequate mathematical models of objects and perturbations acting on them in real conditions makes it necessary to use full-scale mathematical modeling using realizations of full-scale data and special recalculation models in the "small" [7] or small-scale physical models operating in "accelerated" time with observance of similarity conditions of model and full-scale systems

5. Diagram of simulated full-scale mathematical modeling of the object control system with a purposefully variable structure

As an example of simulated full-scale mathematical modeling the diagram in figure 3 can be considered, in which, in addition to the notations of figure 1 the following is added: Z – is the cost of production; Q – performance indicators; indices "M" – model; "FSM" –full-scale model.

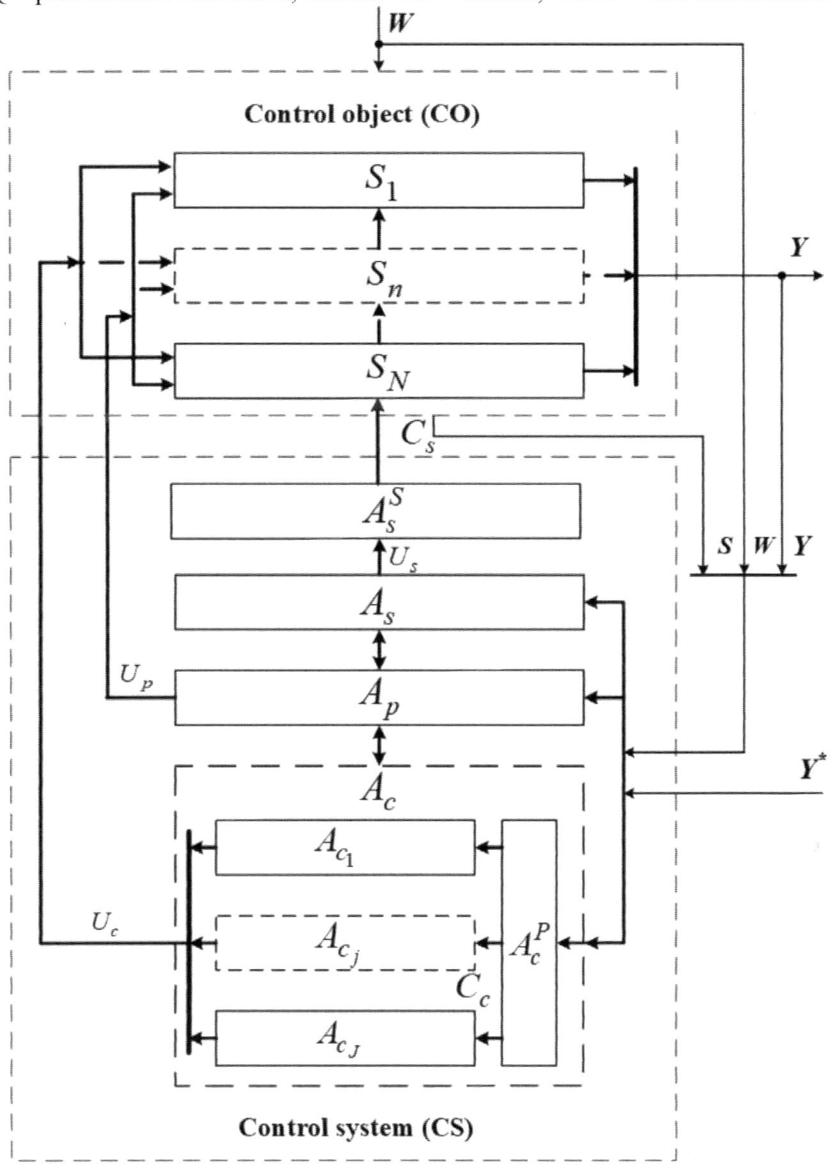

Figure 1. Generalized diagram of the object control system with a purposefully variable structure.

S_n – the n-th structure of the CO; $n=\overline{1,N}$; A_s, A_p, A_c – algorithms of structural, parametric and coordinate control; A_{cj} – the j-th structure of the coordinate control algorithm A_c, $j=\overline{1,J}$; A_s^S, A_c^P – algorithms for switching the CO structure of and the algorithm A_c; C_s, C_c – control commands to switch structures; W and Y, S – vectors of external and output actions, states of the object; Y^* – task on Y, targets, constraints; U_s, U_p, U_c – structural, parametric and coordinate control actions.

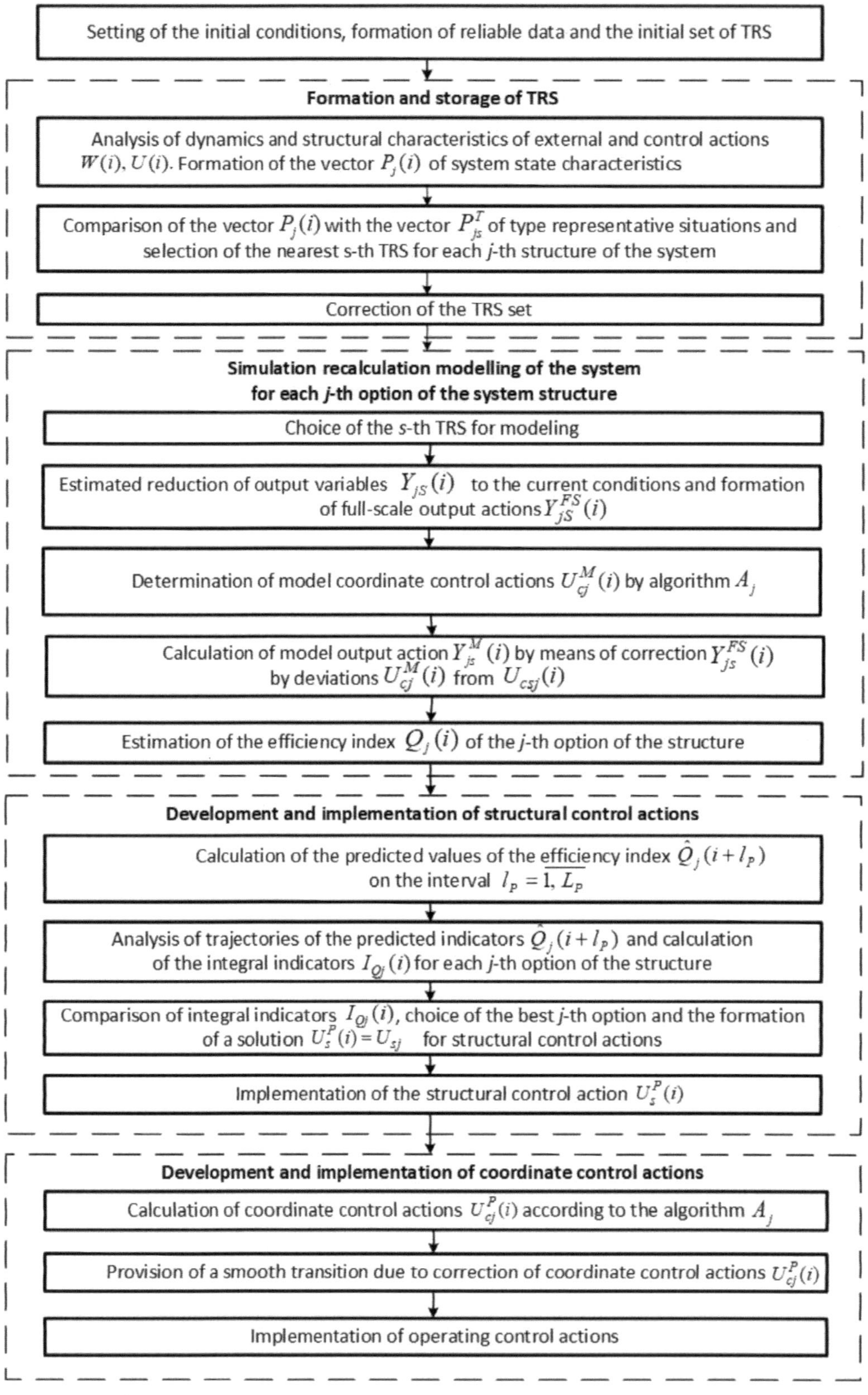

Figure 2. Algorithm for object control with a purposefully variable structure.

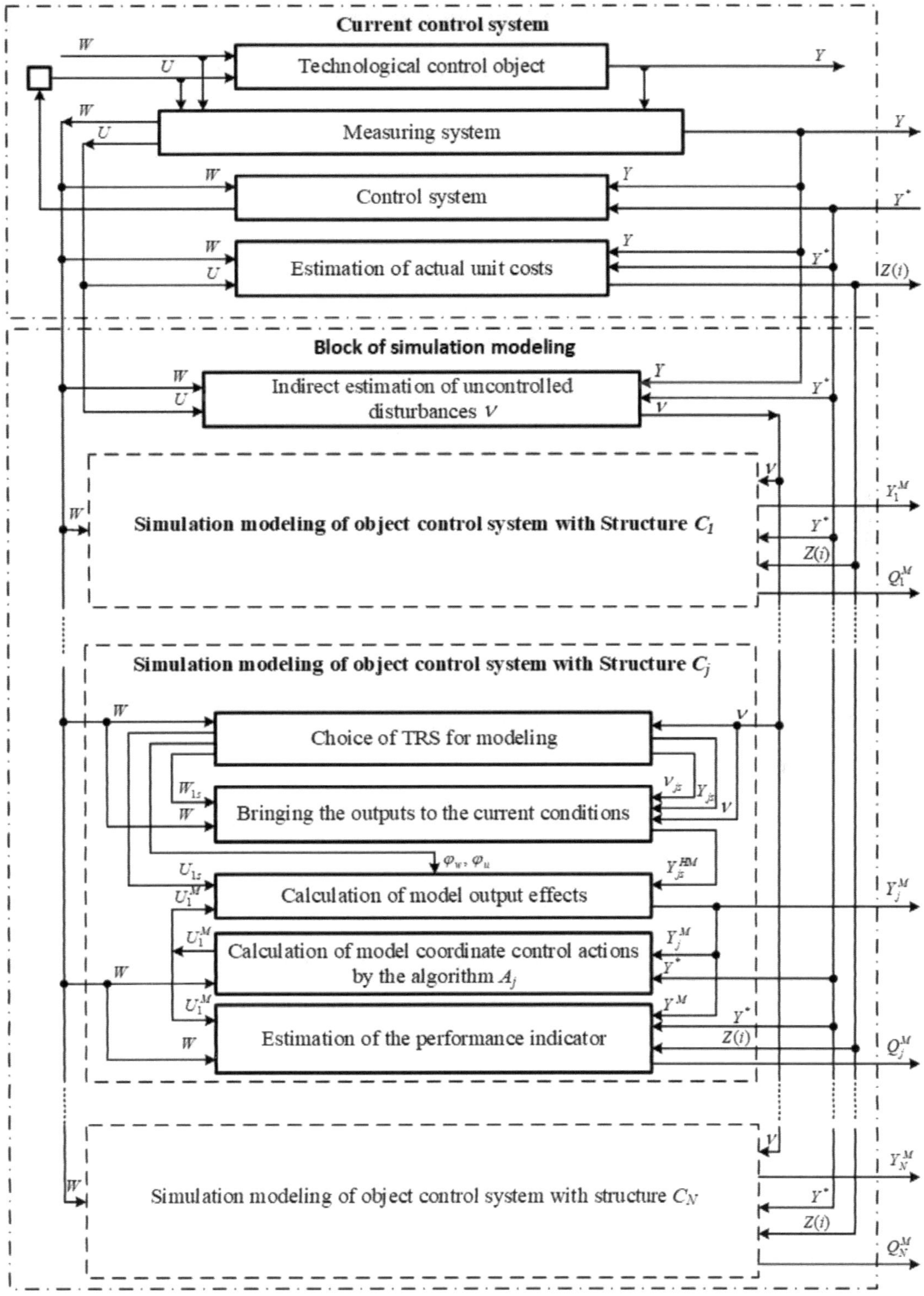

Figure 3. Diagram of simulation full-scale mathematical modeling of the object control system with a purposefully variable structure.

IOP Conf. Series: Materials Science and Engineering **354** (2018) 012001 doi:10.1088/1757-899X/354/1/012001

6. Conclusion

The refinement of the proposed approaches, the results of their implementation at the mining enterprises gave grounds for drawing a conclusion about their effectiveness and the need to expand theoretical research and to disseminate across technical, social and economic sectors.

Acknowledgements

The work was carried according to the State Assignment of the Ministry of Education and Science of the Russian Federation No. 8.8611.2017 / 8.9.

References

[1] Emelyanov S V and Nappelbaum E L 1981 *System Research: Yearbook* (Moscow: Nauka) pp 7–86

[2] Emelyanov S V, Korovin S K, Rykov A S et al. 2007 *Methods of Iidentification of Industrial Facilities in Control Systems* (Kemerovo: Kuzbassvuzizdat) p 307

[3] Emelyanov S V, Korovin S K, Myshlyaev L P et al. 2008 *Theory and Practice of Prediction in Control Systems* (Kemerovo: Russian Universities, Kuzbassvuzizdat – ASTSh) p 487

[4] Emelyanov S V 1967 *Automatic Control Systems with Variable Structure* (Moscow: Nauka) p 336

[5] Emelyanov S V and Korovin S K 1997 *New Types of Feedback: Control under Uncertainty* (Moscow: Nauka. Fizmatlit) p 352

[6] *Industrial Automation Systems* 2006 ed L P Myshlyaev vol 2 (Moscow: Nauka) p 483

[7] Avdeev V P, Kustov B A and Myshlyaev L P *Production and Research Systems with a Multi-variant Structure* 1992 (Novokuznetsk: Kuzbass branch of the Engineer Academy) p 188

[8] Evtushenko V F, Burkov V N, Myshlyaev L P and Makarov G V 2017 *IOP Conference Series: Earth and Environment Science* **84** 1–5

[9] Evtushenko V F, Burkov V N, Myshlyaev L P and Makarov G V 2017 *IOP Conference Series: Earth and Environment Science* **84** 1–5

IOP Conf. Series: Materials Science and Engineering **354** (2018) 012002 doi:10.1088/1757-899X/354/1/012002

Precedent approach to the formation of programs for cyclic objects control

Kulakov S M, Trofimov V B, Dobrynin A S and Taraborina E N

Siberian State Industrial University, 42 Kirova street, Novokuznetsk, 654007, Russia

E-mail: kulakov-ais@mail.ru

Abstract. The idea and procedure for formalizing the precedent method of formation of complex control solutions (complex control programs) is discussed with respect to technological or organizational objects, the operation of which is organized cyclically. A typical functional structure of the system of precedent control by complex technological unit is developed, including a subsystem of retrospective optimization of actually implemented control programs. As an example, the problem of constructing replaceable planograms for the operation of the link of a heading-and-winning machine on the basis of precedents is considered.

1. Introduction

The functioning of a large number of objects (mechanisms, machines, vehicles, technological units, man-machine complexes of different scale and purpose) as well as the work of small and large groups of people is organized cyclically. The latter implies the repetition, with some accuracy, of a certain set of actions, controls, operations, the implementation of which allows the desired result to be obtained at the end of each work cycle. Multiple playback and control of cycles for a certain period of time (shift, day, month) allows the data on control programs, conditions and results of their implementation, on the trajectories of change in controlled variables to be accumulated. These data should be used to form effective programs for controlling the upcoming work cycles of the facility in conditions of change in the properties of disturbing influences and control channels, as well as the diversity of cyclic assignments (orders) for the products received.

The paper presents: the task of creating and implementing cyclic programs for controlling a complex technological object (CTO) or a production complex, a new structure of a program control system, a procedure (algorithm) for the formation of a program for each upcoming work cycle of an object. A distinctive feature of the proposed system and procedure is a new interpretation and concretization of the well-known in the international legal practice method of decision-making based on precedents [1], as well as a similar method of selecting regular (recurring) decisions in intelligent control systems based on precedents [2-5], which are combined with the idea of retrospective optimization and control prediction [6, 7].

2. The precedent method of decision-making

Before presenting the essence and specifics of the proposed precedent approach to program control of cyclic objects, we need to turn to the terminology and content of the well-known method of judicial decision-making based on precedents and to the methods of reasoning based on precedents relating to the theory of intellectual systems.

Content from this work may be used under the terms of the Creative Commons Attribution 3.0 licence. Any further distribution of this work must maintain attribution to the author(s) and the title of the work, journal citation and DOI.
Published under licence by IOP Publishing Ltd

Precedent (*praecedens*, lat. – going ahead) is a case that had an earlier place and serves as an example or justification of cases of this kind [8].

Legal precedent is the decision of a certain court or executive authority on a specific judicial or administrative case, which has become a model for solving similar cases in the future [9].

Case law – the legal system (England, the United States, Canada, and other countries), in which the main source of law is the *judicial precedent*. In the absence of clear definitions of the law, judges have the power and duty to create the law by setting a precedent on the basis of "similar cases". The decision on the particular case in question will be applied in the future to other similar cases.

According to [2, 8, 9] reasoning methods based on precedents include 4 main stages of decision making, which constitute a *precedent cycle*:

- Retrieve [*ri'triv*] – extract from the case library a similar precedent or a subset of precedents, corresponding to the situation in which another (new) decision is need to be made.
- Reuse [*ri:'ju:z*] – reuse of the extracted precedent (precedents) for making the next decision.
- Revise [*ri'vaiz*] – revision (correction), if necessary, of a precedent decision in accordance with the considered (next) task (problem).
- Retain [*ri'tain*] – saving the newly adopted solution as part of a new precedent (for future similar situations).

The structure of the precedent cycle is shown in figure 1.

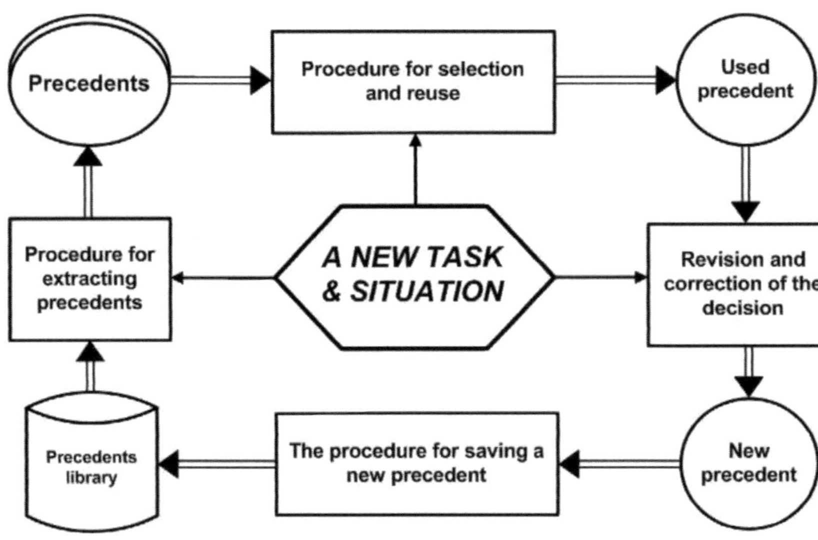

Figure 1. General presentation of the precedent decision-making cycle in similar situations.

The model of each precedent includes the description of three main components: the situation, the solution of the problem, the result of applying the solution. In the cases when such description is not sufficient for the implementation of the precedent cycle, the ontology (formal description) of the subject area, to which the decision problem belongs, is used. Ontology, in turn, includes: a finite set of concepts (notions, terms), a finite set of relations between concepts, a finite set of interpretation functions defined on concepts and relationships.

The tasks of operational control and scheduling of complex technological units and cyclic-class complexes are usually formulated and solved on the basis of functional, physico-chemical, balance and other models of control objects. The examples of such systems are modern automatic control systems of technological processes (ACS TP) for steelmaking units (converters, arc furnaces), chamber heating and thermal furnaces, coke batteries, rolling mills, production of tank structures, complex machine tools and vehicles, coal mining complexes in mines, etc. Their disadvantage is the presence of significant errors in control actions, which are due to the inaccuracy of the models used, incompleteness of information about changes in the properties of the control object and multivariate perturbation effects. Such ACS TP do not take into account the fact that each cycle of the object operation generates not only material products (technical products, materials, energy), but also data sets that represent the information model of a particular technological cycle.

3. The task of precedent control of a complex technological object

The above precedent approach to the selection of control decisions (effects), in the form of a control program for each upcoming technological cycle, allows the operational information about the operation of the object and the control system to be fully used and, as a result, the number of control errors to be reduced (its efficiency to be increased). Next, let's look at how the functional-algorithmic structure of the control system of a cyclic action unit should look, which is based on the precedent approach to the development of control programs for regular (repeated) cycles.

When developing automated control procedures for CTO, it is important to consider that each technological cycle is often not an exact repetition of previously implemented cycles. The latter are more or less different: the goals of control (the properties and characteristics of the product), the changing properties and parameters of the process unit (because of its aging, the use of different materials and regimes, load changes), the experience and control techniques of different operators – technologists, properties of external (disturbing) effects. The precedent cycle of the formation and implementation of control programs for such objects, presented in figure 1, should be changed and supplemented to take into account the named features and distinctive features.

The task of precedent control of CTO can be formulated as follows.

Given: a) A meaningful description of the existing complex technological object in accordance with figure 2.

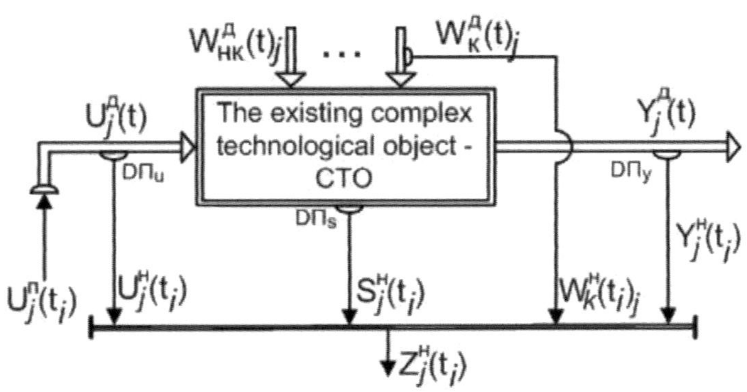

Figure 2. General representation of the control object – a complex technological unit (a complex).

$Y_j^R(t), U_j^R(t), W_C^R(t)_j, W_{NC}^R(t)_j$ – the real CTO output actions on the external environment in the j-th technological cycle at time t, the real control actions on the CTO, the real controllable external actions on the CTO, the real non-controllable external action on the CTO, respectively; $Y_j^{FS}(t_i), U_j^{FS}(t_i), W_C^{FS}(t_i)_j, S_j^{FS}(t_i)$ – a vector of full-scale measuring signals (data) about the output actions of the object related to the j-th technological cycle and the discrete moment ↑↓ of time t_i, the vector of signals about the full-scale control actions on the object, the vector of full-scale signals of controlled external actions (disturbances), the vector of full-scale signals characterizing state of the object in the j-th cycle at the time t_i; $Z_j^{FS}(t_i)$ – generalized vector of full-scale signals received from

IOP Conf. Series: Materials Science and Engineering **354** (2018) 012002 doi:10.1088/1757-899X/354/1/012002

sensors-converters; SC_U, SC_S, SC_Y, SC_W – conventional designation of actuators (A) and sensor-converters (SC).

b) Retrospective information model of the functioning of the existing control system in the form of a database on the parameters $\{Z^N_{j-m}(t_i); m=1,2,...,M; t_i(j-m) \in T_{j-m}\}$ of the actual realized technological cycles.

c) Difference (recalculation) mathematical models of the influence of increments $\Delta u_p(t_i) = u_p^{FS}(t_i) - u_p(t_i)$ of control actions $u_p(t_i) \in U_j(t_i), p \in P_j$ and controllable disturbances $\Delta w_{cr}(t_i) = w_{cr}^{FS}(t_i) - w_{cr}(t_i)$ on changes in output variables $\Delta y_n^A(T_j)$ and states $\Delta s_m^A(T_j)$ of the object at the end of the technological cycle.

d) Criterion of optimality of technological cycle control T_j, for example, the total costs $Q(T_j)$ on the cycle realization.

e) Limitations on control actions of $U_j(t_i) \in \Omega(U, T_j)$ type, where $\Omega(U, T_j)$ is the range of admissible values of the control vector components in the cycle T_j.

f) Information on the starting state $\tilde{S}(t_0)_{j+1}$ of the control object at the time t_0 of the beginning of the next (j+1)-th period (cycle) of its operation, as well as on controllable disturbances $\tilde{W}_C(t_0)_{j+1}$.

Required: to modify the method of decision-making on the basis of precedents applied to the control of CTO of cyclic action, to specify it in the form of a procedure (algorithm) for the formation and implementation of control programs and the functional structure of the precedent control system.

The effectiveness of the precedent method with regard to the control of the CTO of a cyclic action largely depends on the number of high-quality and relevant precedents (corresponding to the current state of the object, external influences and the task of the upcoming technological cycle, as well as error-free solutions, operators – technologists in the immediate past cycles). With a considerable variety of tasks for production output (for example, more than 400 grades of steel can be smelted in a modern steel smelting shop), as well as with a large capacity of many situations for external actions and measurement errors, it is very difficult to select at least 5 faultless precedents for each type of a product. Therefore, it is advisable, at the end of each technological cycle, not to reject erroneous decisions, but to correct them, solving the problem of retrospective optimization. It is possible to rely on this optimization in the method of restorative-predictive automatic control described in [6, 7] and protected by a number of patents for inventions.

The selection of precedents from the set of past technological cycles in relation to the forthcoming cycle can be carried out on the basis of coincidence of goals and constraints, the proximity of the quality parameters of the products received, the proximity of the duration of the completed and forthcoming cycles, the closeness of the initial conditions, and the lowest costs for output of the unit of production.

4. Stages of formation of control programs based on precedents

The procedure for solving the above problem on the basis of the precedent method includes the following components:

1. Control of parameters $Z_j^{FS}(t_i) = \{Y_j^{FS}(t_i), S_j^{FS}(t_i), W_C^{FS}(t_i)_j, U_j^{FS}(t_i), t_i \in [0, T_j]\}$ and calculate the estimates of the performance indicators $Q_m(T_j) = F_m\{Z_j^{FS}(t_i), t_i \in [0, T_j]\}, m \in M_E$ of the control system in the course and at the end of the j-th time interval $[0, T_j]$. In particular, the cost of implementing the j-th technological cycle can be used as the main indicator of efficiency. The components of the vector $Z_j^{FS}(\cdot)$ have the form:

$$Y_j^{FS}(t_i)=[y_1^{FS}(t_i),...,y_n^{FS}(t_i),...,y_N^{FS}(t_i)]_j; \qquad S_j^{FS}(t_i)=[s_1^{FS}(t_i),...,s_l^{FS}(t_i),...,s_L^{FS}(t_i)]_j$$
$$W_{Cj}^{FS}(t_i)=[w_{C1}^{FS}(t_i),...,w_{Cr}^{FS},....,w_{CR}^{FS}(t_i)]_j; \qquad U_j^{FS}(t_i)=[u_1^{FS}(t_i),...,u_p^{HS}(t_i),...,u_P^{FS}(t_i)]_j \qquad (1)$$

All data on the measured values of the parameters, on the actually implemented control program $PR(j,t_i)$, as well as on estimates of the calculated indicators $\{Q_m^{FS}\cdot(T_j)\}$ are placed in a special database.

2. Retrospective optimization (at the end of the j-th technological cycle) of the actually implemented program:

$$PR^{FS}(j,t_i)=\{u_1^{FS}(t_{1i},\tau_1),....,u_p^{FS}(t_{pi},\tau_p),...,u_P^{FS}(t_i,\tau_P);Y_j^{FS}(t_i);S_j^{FS}(t_i);t_i,t_{pi}\in[0,T_j]\} \qquad (2)$$

where $\{u_1^{FS}(\cdot),....,u_P^{FS}(\cdot);t_i\in[0,T_j]\}$ is the actual program of control actions; $Y_j^H(\cdot)$ – the actual program of output variables of the object; $S_j^{FS}(\cdot)$ – the actual program of variables of the object state; $\tau_1,...,\tau_p,....,\tau_P$ – application time of the 1,...,p,...P-th control actions; t_i – current (i-th) time point belonging to the cycle $[0,T_j]$; t_{pi} – the moment of the beginning of application of the p-th control action. Optimization is carried out in accordance with formula (3):

$$PR^O(j,t_i)=\begin{cases}PR^{FS}(j,t_i),\ if\ Y_j^{FS}(T_j)=Y_j^*(T_j);\ S_j^{FS}(T_j)=S_j^*(T_j);\\ Ag\{PR^{FS}(j,t_i);t_i,t_{pi}\in[0,T_j]\};\ if\ Y_j^{FS}(T_j)\neq Y_j^*(T_j);\ S_j^{FS}(T_j)\neq S_j^*(T_j)\end{cases} \qquad (3)$$

where $PR^O(j,t_i)$ is the optimized control program for the j-th technological cycle; $Ag\{\cdot\}$ – the operator (algorithm) of retrospective optimization [6, 7] of the actual (full-scale) program $PR^{FS}(j,t_i)$ implemented in the period $[0,T_j];Y_j^*(T_j),S_j^*(T_j)$ – the specified values of the output variables and variables of the object state at the end of technological cycle, i.e. at the point of time $t_i=T_j$.

$$PR^O(j,t_i)=\{u_1^O(t_{1i},\tau_1),...,u_P^O(t_i,\tau_P);Y_j^O(t_i);S_j^O(t_i);W_C(j,t_i);t_i,t_{pi}\in[0,T_j]\} \qquad (4)$$

The first line of formula (3) corresponds to the case when the actual and target values of the goal variables $Y_j^{FS}(T_j),S_j^{FS}(T_j),Y_j^*(T_j),S_j^*(T_j)$ (with a given accuracy) almost coincide, that is, the operator $Ag\{\cdot\}=1$. The second line is realized when the specified values of the goal variables are significantly different.

The operator $Ag\{\cdot\}$ uses the difference mathematical model of the control object of the following form:

$$\begin{cases}\Delta y_n^A(T_j)=\Phi_n\{[u_P^{FS}(t_i)-u_P(t_i)]_j;p\in[1,P];n\in[1,N];u_P^{FS},u_P\in U_j^{FS}\},\\ \Delta S_m^A(T_j)=F_m\{[u_P^{FS}(t_i)-u_p(t_i)]_j;m\in[1,M];p\in[1,P];u_p^{FS},u_p\in U_j^{FS}\},\end{cases} \qquad (5)$$

where $\Phi_n\{\cdot\},F_m\{\cdot\}$ the operators of the recalculation (difference) type, which in the simplest form represent the tables of recalculation coefficient:

$$\Delta u_p \to \Delta y_n; \qquad \Delta u_p \to \Delta s_m \tag{6}$$

The optimized control programs $\{PR^O(j,t_i); j \in J; t_i \in [0,T_j]\}$ are placed in a special database of optimal (rational) programs, which are considered as precedents at the stage of forming control programs for the upcoming technological cycles.

3. Formation of a subset of actual precedents for the forthcoming (j+1)-th period of the object's operation.

$$\{P\hat{R}(j-m), m \in [0,M]\} \Leftarrow B\{[Y^o_{j-m}(T), Y^*_{j+1}(T), m = \overline{0,M}]; [S^O_{j-m}(T), S^*_{j+1}, m = \overline{0,M}]\}, \tag{7}$$

where $B\{\cdot\}$ – an algorithm for selecting precedents that correspond to given values of the target variables $Y^*_{j+1}(T), S^*_{j+1}(T)$ of the upcoming (j+1)-th cycle.

Selection of programs $\{P\hat{R}(j-m)\}$ is carried out by the criterion of maximum closeness of values to the given values $y^O_{j-m}(T), s^O_{j-m}(T)$. Elements of the set $\{P\hat{R}(j-m), m \in [0,M]\}$ are ordered in descending order of proximity criterion.

4. Interactive selection by the operator – technologist of the retrospective-optimal program $PR^*(j+1,t_i)$ from an ordered set $\{P\hat{R}(j-m)\}$ that corresponds to his preferences $PR^*(j+1,t_i) \in \{P\hat{R}(j-m), m \in [0,M]\}$.

5. Adjustment of the selected program-precedent $PR^*(j+1,t_i)$ in the presence of significant deviations in the parameters of disturbances $W_C^{FS}(j+1,t_i)$, the upcoming cycle of operation of the technological control object from the corresponding parameters $W_C(j+1,t_i)$ of the selected program. In this case, a difference mathematical model (5) is used, which is supplemented by a difference model of the effect of controlled perturbations.

5. Functional structure of the precedent control system

Figure 3 shows a functional diagram of the control system that implements the above described procedure for formation of programs controlling the complex object based on the precedent approach.

An important question in relation to this scheme is the question of the differences and advantages of this method of CTO control from other methods, in particular a method based on type representatives [12] or a method using a retrospective analysis of analogues and prototypes [13, 14]. We believe that the main difference (and advantage) is the use, in real time, of data on the functioning of the current control system, operational accumulation and selection of the best control practices [15]. In this case, it is not necessary to build and continuously update the mathematical model of a complex control object. It is possible to confine oneself to recalculation coefficients in relation to actually implemented controls and controlled disturbances.

As a concrete example of the use of the presented precedent approach, let us consider the task of constructing a planogram of the work of the heading-and-winning combine KM138 (with CGS 13 combine) during the shift [16,17,18]. The planogram is carried out by a replacement link, which includes 10-11 people (combine operator, 7 miners, 2-3 electricians). Parameters of the coal seam: thickness – 1.8 m, face length –180 m. Operating width of the coal combine – 0.63 meters. The standard speed of advancement of the combine is 5 m/min. The standard maintenance time (for one operating cycle) is 30 minutes.

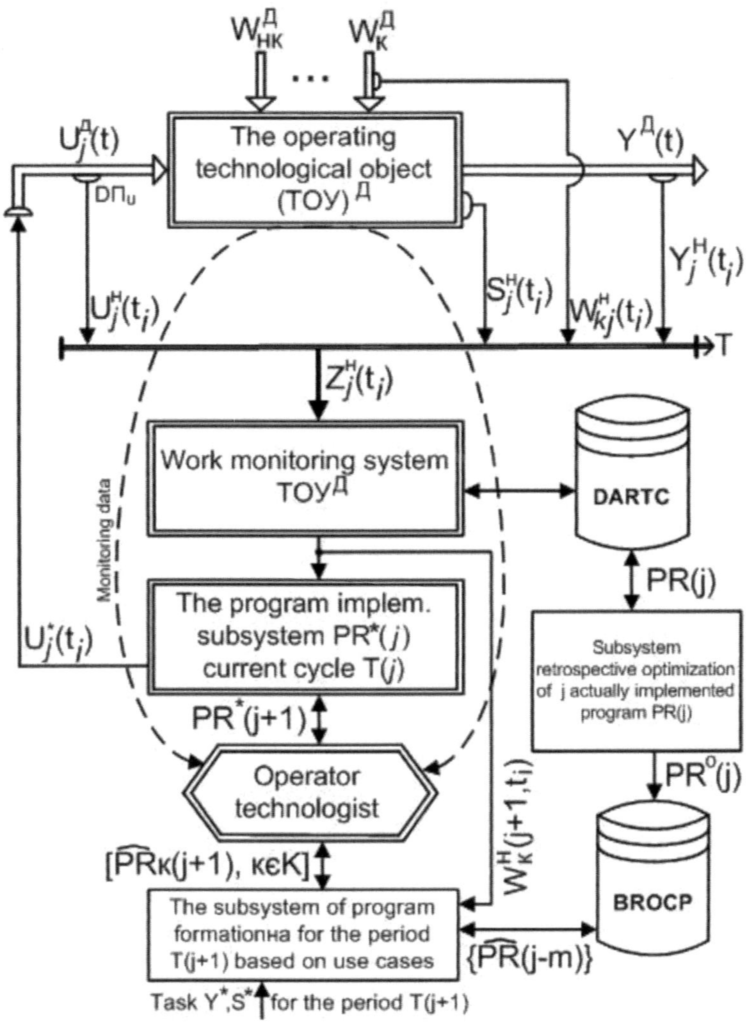

Figure 3. Scheme of formation and implementation of control programs for complex technological objects on the basis of precedents.

(DARTC – database of actually realized technological cycles; BROCP – the base of retrospective-optimal control programs; $PR(j), PR^o(j)$ – actually realized and optimized work program of the object in the period T_j, respectively; $PR^*(j+1)$ – a specified work program for the period T (j + 1); $\{P\widehat{R}(j-m)\}$ – programs from the database of etrospective-optimal control programs for periods $\{T(j), T(j-1),..., T(j-m)\}$.

In each working cycle, the link performs the following typical operations: preparatory and final, cutting of the combine, combine control during coal extraction, auxiliary operations, movement of the support sections, cleaning of the support base, preparation of the combine for hauling, combine control during the hauling, manual cleaning of the coal not loaded with a combine, preparation for moving the conveyor, moving the conveyor, connection of the face with the belt road, connection of the face with the ventilation drift.

IOP Conf. Series: Materials Science and Engineering **354** (2018) 012002 doi:10.1088/1757-899X/354/1/012002

As a planogram – precedent, the schedule shown in figure 4 is selected, which is characterized by accident-free operation during the shift. Its time characteristics are shown in the 1st line of table 1 (a and b).

Table 1. Calculated parameters of the shift cycle.

(a)

No.	Shift starting point t_0	Shift hand-over, min	Start / end of technological cycles, mines				Working speed of the combine, m/min			
			r_1	r_2	r_3	r_4	V_1^*	V_2^*	V_3^*	V_4^*
1	0	38	38 / 125	125 / 210.5	201.5 / 297.5	297.5 / 384.5	5.4	5.0	5.0	5.0
2	0	19.5	19.5 / 104.5	104.5 / 190.0	190.0 / 275	275 / 360	5.5	5.9	5.0	5.0

(b)

Speed of the combine by cycles, m/min				Pauses between cycles, min			Programs
v_1^n	v_2^n	v_3^n	v_4^n	τ_1	τ_2	τ_3	
5.4	5.0	5.0	5.0	7.5x4	7.5x4	6x3	$P\hat{R}_j$
5.5	5.9	5.0	5.0	6.8x4	6.8x4	5.9x4	PR_j^O

The drawback of the precedent is the incomplete completion of the last (4th) cycle of work at the shift end. The lost time was 24.5 minutes. The optimization of the planogram can be performed by reducing time from 38 to 19.5 minutes for reception – shift handover and by 0.1-0.2 minutes of pauses between cycles. The parameters of the optimized planogram are given in the second line of table 1.

It should be noted that the above example of the formation of planograms is illustrative (simplified). A real task of this kind requires the full registration of all processes and actions of the work link during each shift, detailed retrospective analysis and optimization of the planogram, not only by performance criterion, but also by the safety criterion, as well as by other indicators.

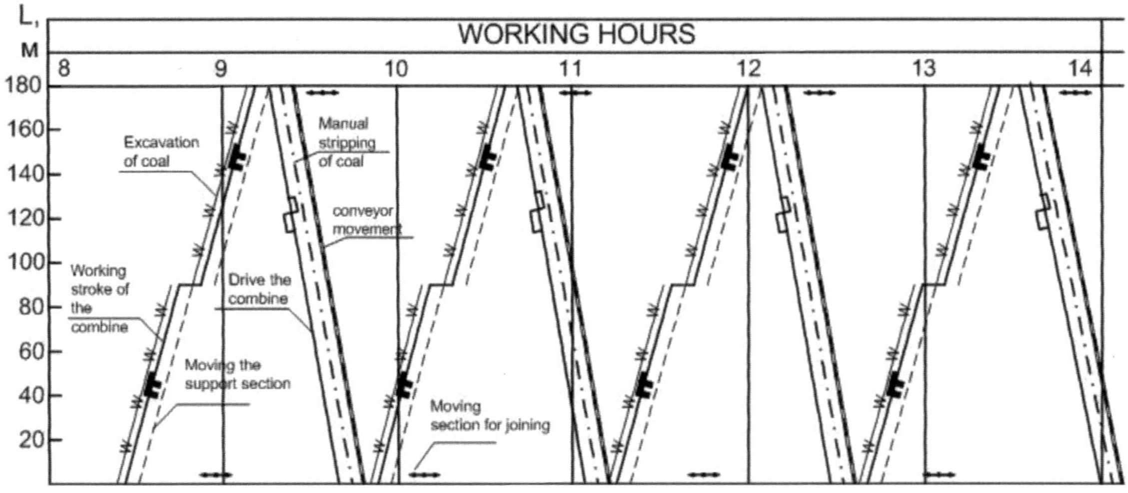

Figure 4. The planogram of the link work, chosen as a precedent.

6. Conclusion

The article considers the problem of applying the precedent approach to decision-making with respect to the construction of management programs for complex technological object of cyclic action. Information on control decisions (control programs) of the operator – technologist, related to the previous technological cycles, is accumulated in the DARTC database. Then, the procedure for retrospective optimization of past decisions that in the BROCP database and, in the future, are used as precedents for choosing a management program for the upcoming technology cycle.

References

[1] Bogdanovskaya I Yu 1993 *Case Law* (Moscow: Nauka) p 239
[2] Varshavsky P R and Alekhin R V 2013 *Information Models and Analyses* vol 2 **4** 386–392
[3] Malykh V L et al 2014 *Information Technologies and Computing Systems* **2** 92–99
[4] Karpov L E and Yudin V N 2007 *Works of the Institute of System Programming of the Russian Academy of Sciences* vol 13 part 2 37–57
[5] Avdeenko T V and Makarova E S 2017 *Bulletin of the State Technical University. Series: Control, Computer Science and Informatics* **3** 85–99
[6] Avdeev V P et al 1984 *Restorative-predictive Control Systems* (Kemerovo: KemSU) p 89
[7] Avdeev V P et al 1978 *Izvestiya Vuzov. Ferrous Metallurgy* **10** 165–168
[8] *A Large Dictionary of Foreign Words* 2003 (Moscow: UNVES) p 784
[9] *Encyclopaedia of the Economist* www.grandars.ru.
[10] Aomond A and Plaza E 1994 *AI Communications* vol **7** **1** 29–59
[11] Varshavsky P R et al 2006 *Artificial Intelligence and Decision Making* **3** 39–62
[12] Avdeev V P et al 1980 *Izvestiya Vuzov. Ferrous Metallurgy* **6** 98–102
[13] Myshlyaev L P et al 2007 *Control Systems and Information Technology* **2.2 (28)** 273–276
[14] Avdeev V P et al 1974 *Izvestiya Vuzov. Ferrous Metallurgy* **10** 163–165
[15] Zimin V V et al 2013 *Fundamentals of Life Cycle Control Services for Informatics and Automation Systems: (Best ITIL Practices)* (Kemerovo: Kuzbassvuzizdat) p 500
[16] Vasyuchenkov Yu F 1990 *Mining. Textbook* (Moscow: Nedra) p 512
[17] Puchkov L A et al 2013 *Underground Mining of Mineral Deposits* (Moscow: Gornaya Kniga) vol 2 p 720
[18] Milekhin G G and Lubin A N 2006 *Processes of Mining Works in the Face* (Murmansk: Apatity branch of Murmansk State Technical University)

Incentive schemes in development of socio-economic systems

V V Grachev[1], K A Ivushkin[2] and L P Myshlyaev[1]

[1]Siberian State Industrial University, 42 Kirova street, Novokuznetsk, 654007, Russia
[2]Consolidated Company "Sibshakhtostroy" LLC, 9 Kuznetskoye highway, Novokuznetsk, 654034, Russia

E-mail: mail@nicsu.ru

Abstract. The paper is devoted to the study of incentive schemes when developing socio-economic systems. The article analyzes the existing incentive schemes. It is established that the traditional incentive mechanisms do not fully take into account the specifics of the creation of each socio-economic system and, as a rule, are difficult to implement. The incentive schemes based on the full-scale simulation approach, which allow the most complete information from the existing projects of creation of socio-economic systems to be extracted, are proposed. The statement of the problem is given, the method and algorithm of the full-scale simulation study of the efficiency of incentive functions is developed. The results of the study are presented. It is shown that the use of quadratic and piecewise linear functions of incentive allows the time and costs for creating social and economic systems to be reduced by 10% -15%.

1. Introduction

The need to stimulate the work executors when creating socio-economic systems is obvious. Timing for the implementation of specific project operations and, in general, the creation of the entire socio-economic system depend on the stimulation of performers quantitatively and sometimes qualitatively.

Until the 90's of the last century, a great number research materials and studies were devoted to stimulation in the construction industry [1-7]. In the works both material and moral forms of stimulation were considered. Some of them [1, 2, 5] mentioned the necessity for stimulation as a means of improving and accelerating the construction industry without specific methodological recommendations. Others considered practical ways of stimulating [3, 4, 7], focusing, as a rule, on research methods of moral effect, for example, socialist competitions.

However, the transition to a market economy, a change in the participants in the construction industry and the demands placed on them required additional study of issues related to the stimulation of construction production as a factor of its acceleration. In the late twentieth century, within the framework of the science "Project Management" (a section of the theory in management of socio-economic systems), the study of incentive mechanisms was carried out in the theory of active systems, in the theory of hierarchical games, in contract theory [8-14].

2. Analysis of previous studies

The greatest results in the development of the theoretical foundations of incentive mechanisms in Russia were achieved by the researchers of the Institute of Problems of Management of the Russian Academy of Sciences (Moscow): Burkov V.N., Novikov D.A., Tsvetkov A.V., Kochiyeva T.B., Zalozhnev A.Yu. and others.

Content from this work may be used under the terms of the Creative Commons Attribution 3.0 licence. Any further distribution of this work must maintain attribution to the author(s) and the title of the work, journal citation and DOI.
Published under licence by IOP Publishing Ltd

In [11, 15-16] four basic incentive schemes are distinguished: jump-like (C-type), compensatory (K-type), proportional (L-type), incentive schemes based on income redistribution (D-type).

Jump-like incentive schemes $\delta_C(x, y)$ (C-type) [15] are characterized by the fact that the agent (performer of operations) receives a permanent reward (equal to the pre-set value of C), provided that the chosen action y is not less than the specified x, and zero reward – if less actions are selected (figure 1):

$$\delta_C(x, y) = \begin{cases} C, & y \geq x; \\ 0, & y < x. \end{cases} \quad (1)$$

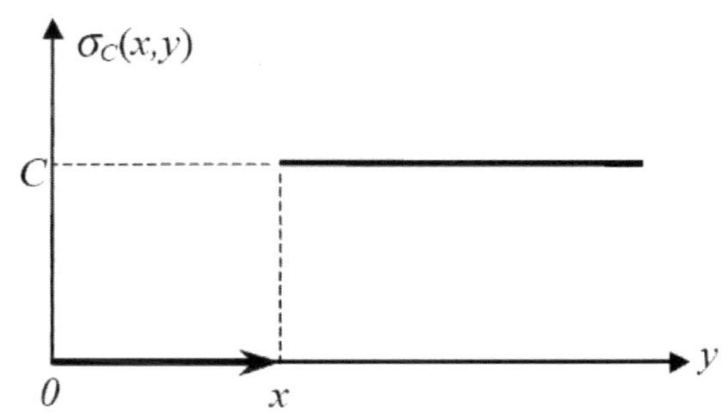

Figure 1. Jump-like incentive scheme (C-type).

Compensatory incentive scheme $\delta_K(x, y)$ (K-type) is characterized by the fact that the agent is compensated for costs, provided that his actions lie in a certain range, given, for example, by restrictions on the absolute amount of individual compensation:

$$\delta_K(x, y) = \begin{cases} c(y), & y \leq x; \\ 0, & y > x; \end{cases} \quad (2)$$

where $x \leq c^{-1}(C), c^{-1}(\cdot)$ is the function inverse to the cost function of the agent, that is, the center can compensate the agent for costs if $y \leq x$ and not pay for the choice of large actions (figure 2).

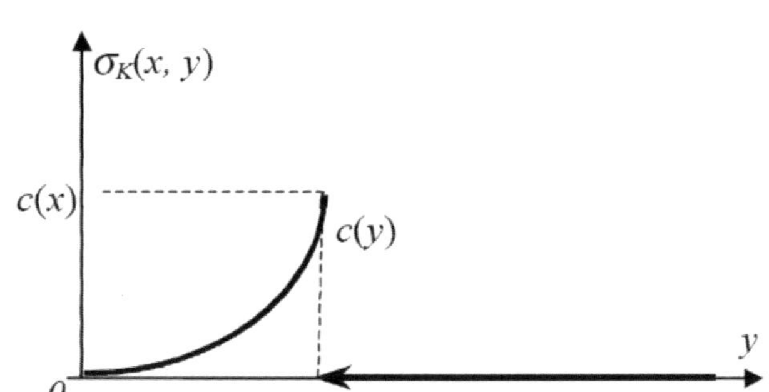

Figure 2. Compensatory incentive scheme (K-type).

Proportional incentive schemes $\delta_L(y)$ (L-type). In practice, wage systems based on the use of constant rates of payment are widespread: time-based payment implies the existence of a rate of payment for a unit of working time (usually an hour or a day), piece-rate payment – the existence of a rate of payment per unit of output, etc. These payment systems are combined by the fact that the remuneration of an agent is directly proportional to its effect (the number of hours worked, the volume of output, etc.), and the payment rate α >0 is the proportionality coefficient (figure 3):

$$\delta_L(y) = \alpha \cdot y \qquad (3)$$

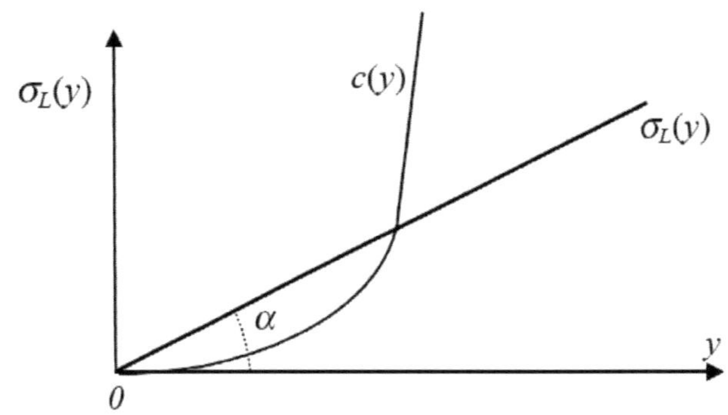

Figure 3. Proportional incentive scheme (L-type).

Incentive schemes based on income redistribution $\delta_D(y)$ (D-type) use the following idea. Since the center expresses the interests of the system as a whole, it is possible to conditionally identify its income and income from the activities of the entire organizational system. Therefore, it is possible to base the agent's incentive on the amount of the center's revenue – set the agent's reward equal to a certain (for example, constant) share of the center's income:

$$\delta_D(y) = \xi \cdot H(y), \qquad (4)$$

where $H(y)$ is the function of the center's income, $\xi \cdot \in$ [0; 1].

Based on the four basic incentive schemes considered above, the systems derived from them are formed, for example CL-type, QD-type.

"The task of stimulation is to choose the optimal incentive scheme that has the maximum efficiency" [15, p. 15]. However, considering various incentive schemes, the authors of the works do not describe methods, algorithms for determining the optimal stimulation system. In works [11, 15-16], the researchers take a priori some stimulation systems, "guess" them, resorting to heuristic procedures, without evaluating the formalized effectiveness of structures of stimulating functions. "One can guess the optimal stimulation system based on substantive considerations, and then correctly justify its optimality" [15, p. 21].

The considered basic incentive schemes – C, K, L, D-type, in our opinion, in practice will not give the proper result. For example, a C-type incentive approach does not encourage the agent – contractor to complete the operation in the shortest possible time. The agent, having received compensation for the achieved action – operation performed over period of time $T > T^{Min}$, will not in the future strive to reduce the time of the operation. When stimulating the K-type, the cost function of the agent $c(y)$ is usually unknown. Agents can distort it in various ways, seeking only to obtain the maximum amount of compensation without shortening the duration of the operation to T^{Min}. Proportional incentive schemes – L-type do not ensure the implementation of the decreasing returns law [17]. The essence of this law is that each subsequent day of reduction in the operation duration requires more stimulation

IOP Conf. Series: Materials Science and Engineering **354** (2018) 012003 doi:10.1088/1757-899X/354/1/012003

than the previous one. With proportional incentive schemes, there is no such increase in funding with a reduction in duration of operations – for each day of duration reduction a fixed amount of stimulus is allocated.

Besides, D.A. Novikov, V.N. Burkov, V.A. Pogozhiya [8, 11, 14-15] in their research, when considering the structures of incentive functions, do not pay due attention to the objects, they are not described as such. Using the game approach, the authors analyze the stimulating functions purely analytically, in absolute terms, without using historical information from previous objects. This approach, based only on analytical methods, requires the construction and use of complex models, attracting a cumbersome mathematical apparatus. Analytical methods are very diverse and strongly depend on the model of the system under study. The variety of these methods is caused, on the one hand, by the desire to obtain a solution analytically, since such solution, as a rule, gives a broader idea of the dependence of the efficiency of stimulating functions on parameters and conditions, and, on the other hand, the mathematical difficulties in obtaining solutions that in different special cases are overcome in their own way. It is typical that, especially for complex systems the application of analytical methods is possible only with substantial simplifications. Thus, in the case of application of analytical methods, it is usually possible to obtain the characteristics of some simplified model of the system.

It is necessary to find new ways for solving the effectiveness evaluation of incentive functions with the most complete extraction of information from existing projects. That is possible with the use of the method of full-scale mathematical modeling [18-20].

From this standpoint, the statement of the problem in the general case can be represented in the following form.

3. Statement of the research task

Given: 1. A set of structures of stimulating functions $St_{i,j}(T)$ for performing operations:

a) linear function

$$St_{1,j} = a \cdot T_j + b, \tag{5}$$

b) piecewise linear function

$$St_{2,j} = \begin{cases} a_1 \cdot T_j + b_1, \text{ if } T_j^{\min} \le T_j \le T_{1,j}; \\ a_2 \cdot T_j + b_2, \text{ if } T_{1,j} \le T_j \le T_{2,j}; \\ a_3 \cdot T_j + b_3, \text{ if } T_{2,j} \le T_j \le T_j^{\text{H}}, \end{cases} \tag{6}$$

where $T_{1,j} = \frac{1}{3}(T_j^{\text{H}} - T_j^{\min})$, $T_{2,j} = \frac{2}{3}(T_j^{\text{H}} - T_j^{\min})$,

c) quadratic function

$$St_{3,j} = a \cdot T_j^2 + b \cdot T_j + c, \tag{7}$$

d) inverse function

$$St_{4,j} = \frac{a}{T_j}, \tag{8}$$

where *a, b, c* – the function parameters, T_j – the duration of the *j*-th operation, T_j^{min} – the minimum execution time of the *j*-th operation, T_j^H – the nominal (in particular, actual) execution time of the *j*-th operation.

2. Data on actually implemented projects, including

- network schedule of design and construction;
- nominal value C_j^N and nominal time of performance T_j^N;
- the basic structure of the stimulation function $St_j^B(T)$, in particular the constant, C-type;
- restrictions on the operations duration

$$T_j^{min} \le T_j \le T_j^N \; ; \tag{9}$$

- restrictions on the amount of incentives for performance of operations

$$0 \le St_j \le St_j^{max} \; , \tag{10}$$

where St_j^{max} – the maximum possible stimulation of the project's operation;

- the function structure of income from the operation of the enterprise after its achievement of design performance $V(t)$.

3. Recalculation modeling procedures, including

- full-scale data on the characteristics of completed projects;
- recalculation models that allow the variations of the input data to be recalculated into a change of output target variables;
- operations of formation of initial model and calculated indicators.

4. Criterion of the project effectiveness $Q(\Delta t)$, reflecting the costs and revenue from the creation of a technological complex

$$Q(\Delta t) = V(\Delta t) - St(\Delta t) \; , \tag{11}$$

where Δt – the time of change in the duration of the project performance, $V(\Delta t)$ – the income of the enterprise over time Δt, $St(\Delta t)$ – the costs for incentives to reduce the project duration by Δt.

***Required*:**

1. To develop an algorithm for determining the effectiveness of stimulating functions.

2. To investigate the effectiveness of stimulating functions $St_{i,j}(T)$ from a given set by the criterion $Q(\Delta t)$.

4. Algorithm for solving the problem

To solve the problem on investigation of the effectiveness of stimulating functions an algorithm was developed, the block diagram of which is shown in figure 4.

This algorithm allows the optimal structure of stimulating functions from the point of view of the criterion $Q(\Delta t)$ to be determined.

The algorithm for determining the effectiveness of stimulating functions consists of three aggregated functional blocks: the input block of input data, the block of simulation recalculation modeling and the block of criterion calculation and selection of the best structure of stimulating function.

Below there is a brief description of each of these three aggregated functional blocks of the algorithm.

Block 1. Input of initial data.

The input of data necessary for the operation of the algorithm is carried out.

1. Input of data on the implemented project.

2. Assignment of structures of stimulating functions $St_{i,j}(T),\ i=\overline{1,m}$.

Initial data for the implemented project include:

- network schedule of design and construction;
- the nominal value C_j^B, the nominal time of execution T_j^B of each operation;
- restrictions on the implementation duration $T_j^{\min} \le T_j \le T_j^H$ and on the stimulation $0 \le St_j \le St_j^{\max}$ of each operation;
- function structure of the income from the enterprise operation after
- its achievement of the design performance.

A set of structures of stimulating functions $St_{i,j}(T)$ was formed from a set of elementary functions that satisfy the decreasing returns law and the simplest and most informative ones. As a result of the selection, the following functions were included in the set

- linear (5);
- piecewise-linear (6);
- quadratic (7);
- inverse proportionality (8).

Block 2. Simulated recalculation modeling of the implemented project for each i-th option of the structure $St_{i,j}(T)$.

Consistently the structure of the incentive function from the given set is considered.

For each project operation, the incentive costs are calculated to shorten the duration of each *j*-th operation for time ΔT

$$St_{i,j}(\Delta T)=St^B_j(\Delta T)+f[Str^B(\Delta T)-Str_i(\Delta T)], \tag{12}$$

where f[·] – the operator for recalculating the variations of the parameters of the current *i*-th structure of stimulation function Str_i from the parameters of the basic structure of the stimulation function Str^B, St_j^B – the value of the basic stimulation of the *j*-th operation.

After the calculation $St_{i,j}(\Delta T)$, the set Θ is formed

$$\Theta^i = \{St_{i,j}(\Delta T)\},\ j=\overline{1,n},\ \Delta T = \overline{1, T_j^H - T_j^{\min}} \tag{13}$$

Further, using the search procedure, the block diagram of which is shown in figure 5, the incentive costs for reduction of the project critical path by the value Δt for each *i*-th option of the structure $St_i(\Delta t)$ are determined.

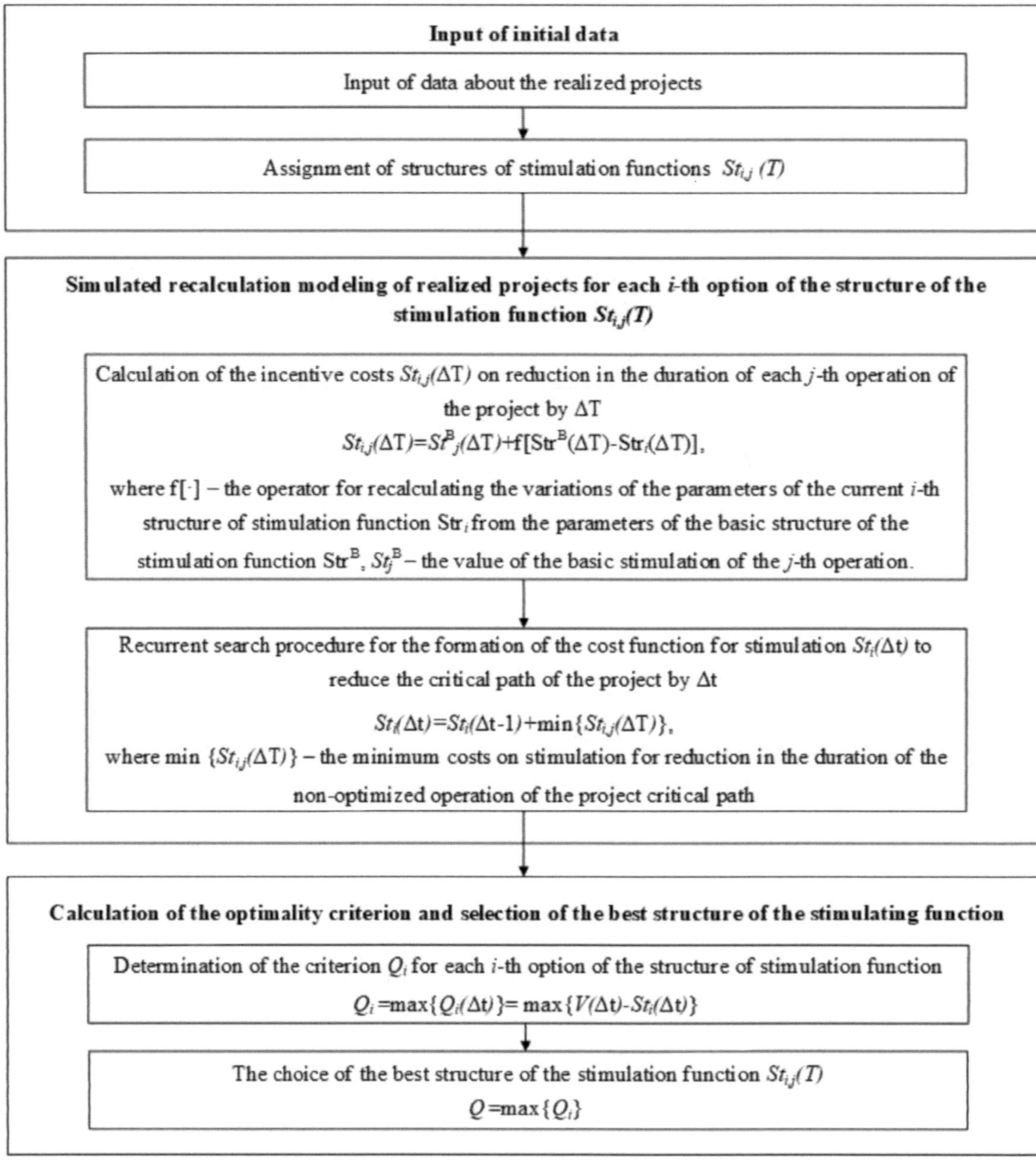

Figure 4. Algorithm for determining the effectiveness of stimulating functions.

The essence of this procedure is to compare the many possible options to reduce the duration of the project critical path by reducing the duration of individual operations and choosing the best from them.

At the first stage, the necessary initial values of the variables are entered for the functioning of the procedure, a set of critical operations θ and the initial length of the critical path T_0^{CP} are determined.

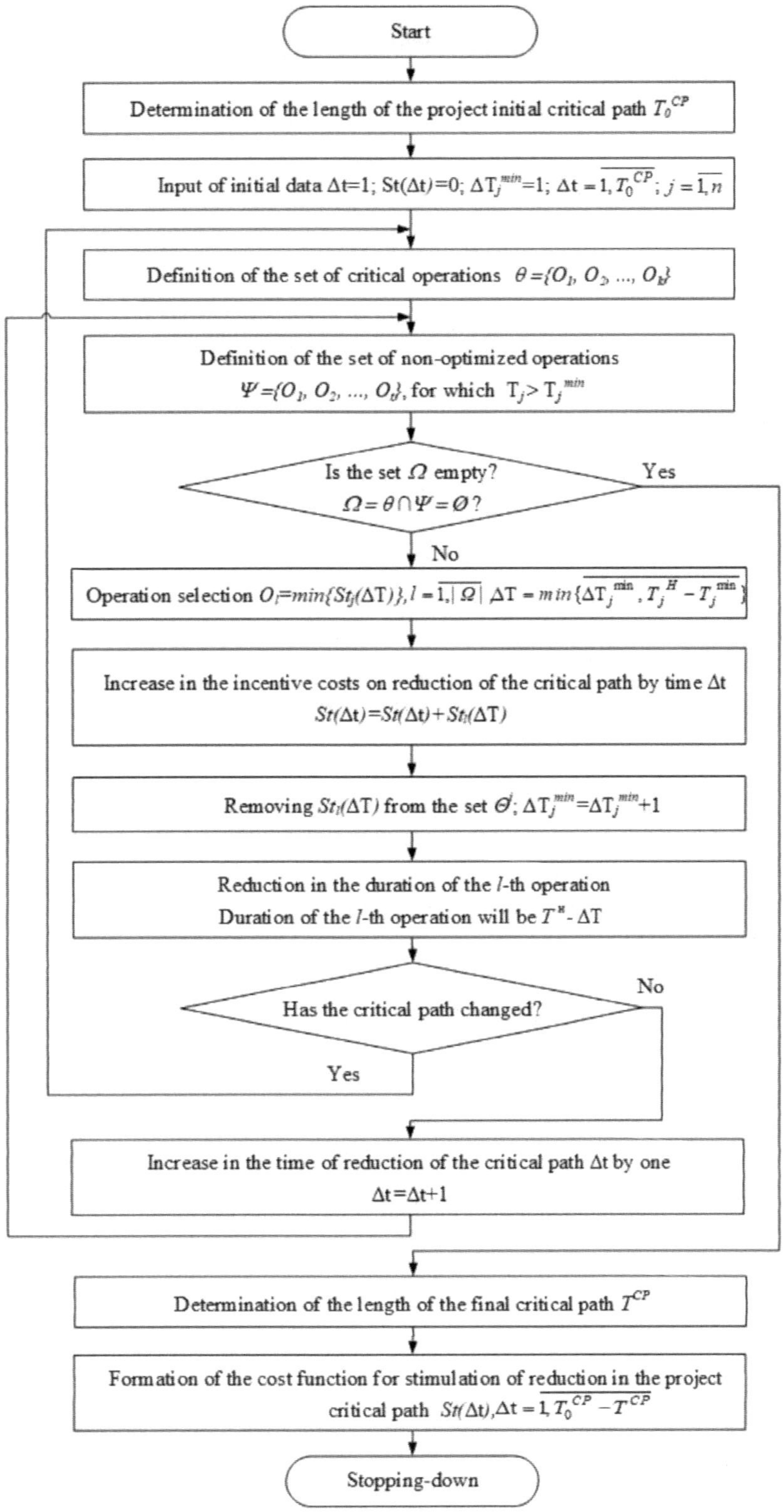

Figure 5. Block diagram of the search procedure for determining the incentive costs on reduction in the project critical path.

Further, from the set of critical non-optimized operations Ω, an operation O_l with minimal incentive costs is found to shorten its duration by time ΔT.

$$O_l = \min\{St_j(\Delta T)\},\ l = \overline{1,|\Omega|},\ \Delta T = \min\overline{\{\Delta T_j^{\min}, T_j^H - T_j^{\min}\}} \quad (14)$$

Then, by the value $St_l(\Delta T)$ we increase the incentive costs on reduction in the critical path by the current time Δt

$$St(\Delta t) = St(\Delta t) + St_l(\Delta T). \quad (15)$$

After we remove $St_l(\Delta T)$ from the set Θ and shorten the duration of the *l*-th operation. After the reduction, the duration of the *l*-th operation will be $T^N - \Delta T$.

We check whether the critical path has changed.

If the critical path of the project has changed, then the described procedure is repeated, while the incentive costs on reduction in the critical path by the current time Δt will constantly increase by the amount of costs on reduction of operations from Θ with the minimum $St_j(\Delta T)$.

If the critical path of the project has not changed, then the incentive costs for reducing the duration of the project critical path $St(\Delta t)$ by the current time Δt are remembered, the current time is increased by one $\Delta t = \Delta t + 1$ and the described search procedure is repeated again. And so long as the possibilities for reduction in the duration of the operations of the set Θ will be exhausted, that is, for the time being, equality $T_j = T_j^{\min}$ will not be fulfilled for all critical works.

As a result of performing the search procedure for each structure from the set of incentive functions, the dependence of the incentive costs on the reduction of the project critical path $St(\Delta t)$, $\Delta t = \overline{1, T_0^{CP} - T^{CP}}$ will be formed, where T_0^{CP}, T^{CP} – the length of the initial and final critical path of the project, respectively.

Block 3. Calculation of the optimality criterion and selection of the best structure of the stimulation function.

For each of the structures of the incentive function $St_{i,j}(T)$, the criterion of effectiveness Q_i is defined

$$Q_i = \max\{Q_i(\Delta t)\} = \max\{V(\Delta t) - St_i(\Delta t)\}. \quad (16)$$

In the final stage, the criterion with the maximum value is selected from the set Q_i

$$Q = \max\{Q_i\}. \quad (17)$$

As a result, the *i*-th structure of the stimulating function will be considered the most efficient, for which condition is fulfilled

$$Q_i = Q. \quad (18)$$

Knowing the best structure of the stimulation function and the greatest value of the optimality criterion Q, to find the optimal time of reduction in the project critical path Δt^{opt} is not a difficult task: at which $Q = \max\{Q(\Delta t^{opt})\}$.

The software realzation of the algorithm is implemented in the Microsoft Project Professional and Microsoft Excel environments in the scripting language Visual Basic for Applications.

The full-scale data for the quantitative factors were taken from the realized project of the coal preparation plant.

5. Conclusions

The results of calculations for the selected set of stimulation functions are shown in figure 6, from which it follows that the best are the quadratic and piecewise linear stimulation functions.

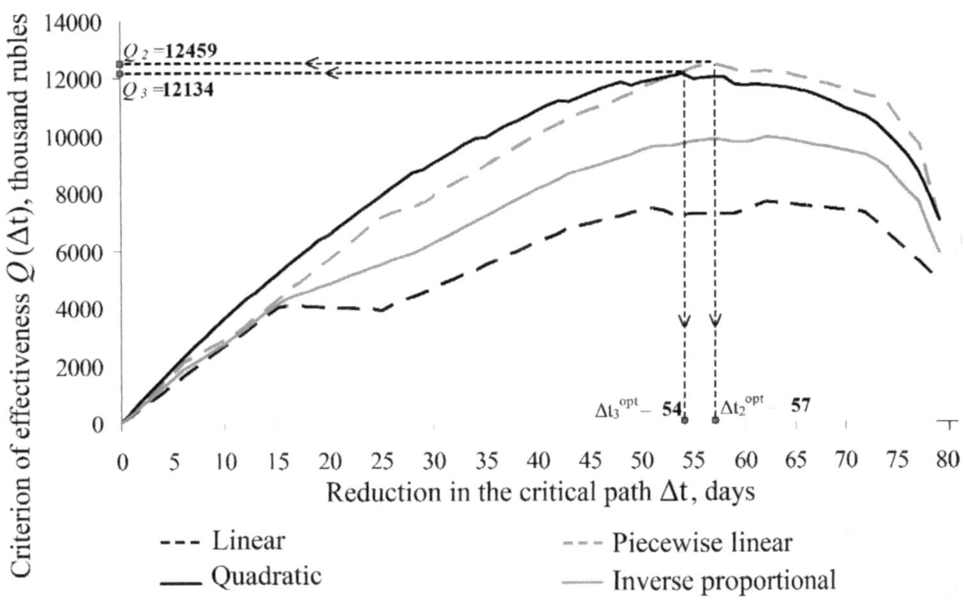

Figure 6. Performance criterion for various structures of incentive function with a company's daily income of 400 thousand rubles.

For the piecewise-linear stimulation function for a specific coal-processing plant, an optimal reduction in the duration of Δt^{opt} was obtained for 57 days at a value of Q equal to 12459.12 thousand rubles, and for a quadratic function – $\Delta t^{opt} = 54$ days, $Q = 12134.47$ thousand rubles.

When the daily income of the enterprise changes, the quadratic and piecewise linear stimulation functions also remain preferable and outperform other stimulation functions by the efficiency criterion Q by at least 50%.

Using the procedure of the recalculation full-scale mathematical modeling allows the incentive values for project operations to be recalculated, while their duration and duration of the whole project change. The network schedule is also dynamically adjusted.

References

[1] Galkin I G, Kantorer S E and Paraubek G E 1978 *Organization, Planning and Management of Building Production* ed I G Galkin (Moscow: Higher School) p 496

[2] Dikman L G 1982 *Organization, Planning and Management of Building Production* (Moscow: Higher School) p 480

[3] O'Brien D 1971 *Application of the Critical Path Method in Construction* (Moscow: Publishing House of Literature on Construction) p 168
[4] Rybalsky V I 1974 *Automated Production Management Systems* (Kiev: Vishcha Shkola) p 480
[5] Komarov I K 1979 *Perfection of the Construction Industry* (Moscow: Stroyizdat) p 208
[6] Kazansky Yu N 1985 *Experience in the Organization and Management of Construction Firms in the United States* (Moscow: Stroyizdat) p 269
[7] Ushatsky S A 1974 *Choosing the Optimal Solutions in the Management of Construction Production* (Kiev: Budivelnik Publishing House) p 168
[8] Pobozhy V A, Pavlenko S I, Pobozhiya M V and Tkachenko V V 2001 *Calculation and Optimization of Network Construction Sschedules* (Moscow: Publishing House of Construction Universities Associations) p 240
[9] Burkov V N and Novikov D A 1997 *How to Manage Projects* (Moscow: Sinteg) p 188
[10] Kochieva T B and Novikov D A 2000 *Basic Incentive Schemes* (Moscow: Apostrof) p 108
[11] Novikov D A 1998 *Stimulation in Socio-Economic Systems* (Basic Mathematical Models) (Moscow: IPP RAS) p 216
[12] Novikov D A and Tsvetkov A V 2000 *Mechanisms of Stimulation in Multi-element Organizational Systems* (Moscow: Apostrof) p 184
[13] Tsvetkov A V 2001 *Stimulation in Project Management* (Moscow: Apostrof) p 144
[14] Burkov V N, Zaloznev A Yu and Novikov D A 2001 *Graph Theory in the Management of Organizational Systems* (Moscow: Sinteg) p 124
[15] Novikov D A 2003 *Mechanisms of Stimulation in Organizational System*s (Moscow: IPP RAS) p 147
[16] Novikov D A 2005 *Theory of Management of Organizational Systems* (Moscow: Moscow Psychological and Social Institute) p 584
[17] Raizberg B A, Starodubtseva E B and Lozovsky L Sh 1999 *Modern Economic Dictionary* (Moscow: INFRA-M) p 479
[18] Venikov V A and Sovetov V A 1976 *Theory of Similarity and Modeling* (Moscow: Higher School) p 479
[19] Avdeev V P, Zeltser S R, Kartashov V Ya and Kiselev S F 1987 *Naturopathic and Mathematical Modeling in Control Systems* (Kemerovo: KemSU) p 84
[20] Avdeev V P, Kustov B A and Myshlyaev L P 1992 *Production-research Systems with a Multivariant Structure* (Novokuznetsk: Kuzbass branch of the Engineering Academy) p 188

Investigation of multidimensional control systems in the state space and wavelet medium

D B Fedosenkov[1], A A Simikova[2] and B A Fedosenkov[2]

[1]Siberian Generating Company, 53 build. 5, Doubininskaya Street, 115054, Moscow, Russia
[2]Kemerovo State University, 6 Krasnaya Street, Kemerovo, 650000, Russia

E-mail: rafwavelet@ngs.ru

Abstract. The notions are introduced of "one-dimensional-point" and "multidimensional-point" automatic control systems. To demonstrate the joint use of approaches based on the concepts of state space and wavelet transforms, a method for optimal control in a state space medium represented in the form of time-frequency representations (maps), is considered. The computer-aided control system is formed on the basis of the similarity transformation method, which makes it possible to exclude the use of reduced state variable observers. 1D-material flow signals formed by primary transducers are converted by means of wavelet transformations into multidimensional concentrated-at-a point variables in the form of time-frequency distributions of Cohen's class. The algorithm for synthesizing a stationary controller for feeding processes is given here. The conclusion is made that the formation of an optimal control law with time-frequency distributions available contributes to the improvement of transient processes quality in feeding subsystems and the mixing unit. Confirming the efficiency of the method presented is illustrated by an example of the current registration of material flows in the multi-feeding unit.The first section in your paper.

1. Introduction

With bringing the concept of "wavelets" into the sphere of automatic control, a need arose to introduce and separate the notions "one-dimensional-point" and "multidimensional-point" concerning automatic control systems (ACS's). In the first - at each physical point of the system a one-dimensional (1D-) signal is considered, in the ACS's of the second type - a multidimensional one (MD-signal). In this case, both these ACS's can be called and considered as multidimensional systems, but putting different semantics into these concepts: 1) a one-dimensional-point ACS from the point of view of the state space method as the basic apparatus of the control theory within the modern paradigm is a multidimensional, or vector, system; the signal vectors (of control $u(t)_i$, state $x(t)_j$, and output $y(t)_k$) are formed in such systems as sets of scalar (1D-) signals distributed over different points of the system; 2) a multidimensional-point system is multidimensional one in its genetics of the representations (i.e., MD-images proper) for 1D-signals, by which the latter are replaced in the system; the signal vectors are formed as vectors physically concentrated at a corresponding point, and they are interpreted as vector-concentrated signals.

2. Optimal systems - internal features

To demonstrate the joint use of approaches based on the concepts of state space and wavelet transforms, a method for optimal control of various technological objects in a state space medium

Content from this work may be used under the terms of the Creative Commons Attribution 3.0 licence. Any further distribution of this work must maintain attribution to the author(s) and the title of the work, journal citation and DOI.
Published under licence by IOP Publishing Ltd

IOP Conf. Series: Materials Science and Engineering **354** (2018) 012004 doi:10.1088/1757-899X/354/1/012004

represented in the form of time-frequency representations (maps), is considered. The technological object analyzed is a mixture-producing aggregate for the production of disperse mixtures.

The computer-aided control system is formed on the basis of the similarity transformation method, which makes it possible to exclude the use of reduced state variable observers.

In the system presented here, the so-called Cohen class distributions [1] are used – to indicate with their help the signals of current material flows. This feature makes the control system effective, and from the user's point of view – more informative and semantically clear.

We consider the linear stationary system of multi-ingredient dosing in the pertinent block of the mixture-producing aggregate, the dynamics of which is described by the equations

$$x(t)=Ax(t)+Bu(t),$$
$$y(t)=Cx(t)+Du(t), \qquad x(t_0)=x_0$$

The system is strictly causal [2], therefore the over-control output matrix D is equal to zero. Here, $x(t)\in R^n$ is the vector of state variables (VSV); $u(t)\in R^m$ – the control vector; $y(t)\in R^r$ – the output vector; $x_0\in R^n$ – the VSV at time $t = t_0$; A, B, C are matrices of dimensions $(n\times n)$, $(n\times m)$, $(r\times n)$.

The problem is to transfer the multi-dosing system from the initial state x_0 to the lowest possible state in a temporal interval $[t_0, t_f]$, without the cost of unduly large control actions $u(t)$ to achieve this goal. In this case, the problem of the optimal regulator (linear quadratic regulator) is formulated as the problem of determining the optimal control law $u^*(t)$ on the interval $[t_0, t_f]$, within which the following optimal criterion J must be minimized:

$$min\leftarrow J\{x(t_0),x(t_f)\}=\int_{t_0}^{t_f}[x^T(t)\cdot Q_x\cdot x(t)+u^T(t)\cdot R\cdot u(t)]dt+x^T(t_f)\cdot Q_f\cdot x(t_f),$$

where $Q_x\in R^{n\times n}$ and $Q_f\in R^{n\times n}$ are symmetric non-negatively definite matrices (respectively, the matrices of weighted integral quadratic quality estimates for the current (Q_x) and final (Q_f) state vectors of the dosing process, and $R\in R^{m\times m}$ is a symmetric positively definite matrix reflecting the contribution of each scalar component of the control vector $u(t)$ to the optimal criterion. In other words, here the so-called Boltz problem [2] is being solved, that is, in fact, a solution is given to the problem of a purposeful change in an optimal control law in response to a detectable effect - during the ongoing monitoring - of changing the complex of characteristics in the material flow at the output of the *j-th* feeder, or at the output of the multi-dosing block. The inclusion of the unit for estimating the unevenness of the material flow in the mixture-producing aggregate into the structure of the optimal control system will enable to realize the optimal control of mixture-producing processes in the state-space medium using current monitoring in a wavelet [3] environment.

In this case, 1D-material flow signals formed by primary transducers (strain gage and piezoelectric sensors) are converted by means of wavelet transformations into multidimensional concentrated-at-a point variables in the form of time-frequency distributions of Cohen's class [1]. Thus, the control law for the total state vector written in the format of the mentioned distributions will look like this:

$$u(t) = K\cdot Coh(t,\omega),$$

where dim $Coh(t,\omega) = (n \times 1)$, t is lapsed time, ω is the instantaneous angular frequency of the atoms on the time-frequency distribution (time-frequency map – TFM), K is the state feedback matrix. Consequently, the equations of state and output for causal systems are written accordingly as follows:

$$\partial\{Coh(t,\omega)\}/\partial t=A\cdot Coh(t,\omega)+B\cdot K\cdot Coh(t,\omega)=(A+B\cdot K)\cdot Coh(t,\omega)=H\cdot Coh(t,\omega)$$

$$y(t) = C\cdot Coh(t,\omega)+D\cdot K\cdot Coh(t,\omega)=(C+D\cdot K)\cdot Coh(t,\omega)$$

Taking into account that a number of state variables (SV's) can not be registered with sensors, it was decided to form alternative state variable vectors v based on the similarity transformation method. In this case, the state, control, and over-state output matrices change according to the expressions:

$$A_v=P^{-1}A\cdot P;\;\; B_v= P^{-1}B;$$

$$C_v=C\cdot P;\;\; D_v=D,$$

where P is a transformation matrix. Note that a pertinent alternative vector $Coh(t,\omega)_v$ is formed depending on the possibilities / impossibilities of their actual registering and also a number of regime/constructive parameters of the dosing unit. In particular, the alternative VSV reflecting the dosing process in a multi-ingredient unit including a discrete-type feeder and two continuous-type feeding devices, was formed, and its vector-matrix model analyzed. The scalar components of the alternative vector were generated in the form of linear combinations of the original vector, provided that only the actual signal components were included in the new VSV: the virtual SV's of the original vector that occurred while representing a signal of the discrete-type feeder as a Fourier model, weren't taken into account. Consequently, when using a new alternative VSV in our research and forming a control signal from the full feedback vector over the converted VSV, there was no need for the use of reduced observers. The latter greatly simplified the structure of the control system. For example, the process of feeding in a multi-component unit was described by a pertinent 17th-order vector-matrix model, and the procedure of inversing the transformation matrix P to turn an initial state vector into a relevant alternative one was realized in the MathCad package.

The optimization of the stationary feeding system in terms of quadratic integral quality estimates was carried out in accordance with the optimality criterion written as the following quadratic form $CohT(t,\omega)\cdot K\cdot Coh(t,\omega) = J$ (Bellman function), in which the symmetric matrix K is determined while solving a relevant Riccati matrix differential equation; in such a case, the quadratic form with matrix K can be written as the Boltz / Lagrange functional [2].

Thus, under the optimal criterion in the Lagrange form, the functional is as follows (the problem of Lagrange):

$$J = Coh^T(t,\omega)\cdot K\cdot Coh(t,\omega)= \int_{t_0}^{t_f} [Coh^T(t,\omega)\cdot Q\cdot Coh(t,\omega)+ u^T(t)\cdot R\cdot u(t)]dt \tag{1}$$

The integrand consists of two quadratic forms:

1) the quadratic form of the system state $Coh^T(t,\omega)\cdot Q\cdot Coh(t,\omega)$ with positively definite symmetric matrix Q, $dim\ Q = (n\ x\ n)$;

2) the quadratic control form $u^T(t)\cdot R\cdot u(t)$ with a positively definite symmetric matrix R, $dim\ R = (m\ x\ m)$. It is required to find the optimal control with a full state feedback, transferring the system from an arbitrary initial point $Coh(t_0,\omega) = Coh(0)$ to a finite point $Coh(\infty) = 0$ and providing the functional minimum.

3. Algorithm of controller synthesis

The algorithm for synthesizing a stationary linear state controller for feeding processes looks like the next one below.

The following conditions must be specified in the source data:

- the system state equation (or directly matrix A and B for the multi-component feeding unit);
- the objective functional J (i.e., the optimal criterion) in the form of equation (1).

The functional J and the two integrands as the quadratic forms are scalar functions.

1. Determine the matrix Q from the quadratic form $Coh^T(t,\omega)\cdot Q\cdot Coh(t,\omega)$ by performing the operations of multiplying the matrix Q with the Cohen's distribution vectors and comparing the result obtained with the form of the given functional J.

2. Similarly, we determine the weight matrix R in the quadratic control form.

3. The next step is the calculation of the elements of the matrix K that is part of the quadratic form, the Bellman function, $J = Coh^T(t,\omega)\cdot K\cdot Coh(t,\omega)$, from the algebraic Riccati equation [2]

$$-K\cdot A - A^T\cdot K + K\cdot B\cdot R^{-1}\cdot B^T\cdot K - Q = 0.$$

Performing the operations of matrix multiplication and, ultimately, addition operations, yields an equation of the form

$$\begin{bmatrix} z_{11} \ldots z_{1n} \\ \ldots\ldots \\ z_{n1} \ldots z_{nn} \end{bmatrix} = \begin{bmatrix} 0 \ldots 0 \\ \ldots\ldots \\ 0 \ldots 0 \end{bmatrix},$$

$$dim\ z = (n \times n),\ dim\ 0 = (n \times n).$$

From the last matrix equation we write n^2 uniform equations, and solving them together, we determine the coefficients k_{ij} of the matrix K.

4. We check the elements k_{ij} of the matrix K for positive definiteness by the Sylvester's criterion [2], and, according to the criterion, determine the true values of the coefficients kij.

5. Substituting the found coefficients k_{ij} into the expression

$$u^*(t) = R^{-1}\cdot B^T\cdot K\cdot Coh(t,\omega)$$

yields the optimal control law.

4. Cohen's class distributions - applied aspects

To obtain a graphical representation of the optimal control law $u^*(t)$, it is necessary to determine all SV's in the Cohen distribution format, included in the optimal control law. The SV's are calculated from the scalar state equations with the SV's in the form of time-frequency distributions, taking into account the matrices A and B.

To minimize the functional J, we used (as mentioned above) multidimensional-point variables in the form of time-frequency distributions of Cohen's class $Coh(t,\omega)$ (in our case, the Wigner-Ville [4] and Choi-Williams [1, 5] distributions) .

When calculating the time-frequency distributions of complex multi-component signals, the latter are expressed in the form of discrete comb functions. In this case, in accordance with the sampling theorem [6], the Wigner-Ville distribution of a comb signal is defined as

$$W(t,\omega) = \frac{T}{\pi} \sum_{k=0,\infty} s^*(t - kT)s(t + kT)\exp(-j2\omega kT), \qquad (2)$$

where 1/T is sampling frequency (sample rate); it is selected from the relation $T \leq \pi/2\omega_{max}$, where ω_{max} is the maximum frequency in the signal spectrum. As it can be seen from the expression (2), the discrete Wigner distribution is periodic with period π (since the frequency of the imaginary exponential is equal to 2ω), and not 2π - as in the case of a continuous signal. Hence, the sampling rate should be twice the Nyquist frequency, that is: $\omega_s \geq 2\omega_N = 4\omega_{max}$.

With the comb form of the material flow signal sampled in time

$$\hat{x}(t) = \sum_{n=0,\infty} x(nT)\,\delta(T - nT)$$

and a continuous frequency ω, we obtain a discrete version of the Wigner distribution:

$$\hat{W}(n,\omega) = \sum_{k=\overline{0,\infty}} s^*((n-2k)T)s(kT)\exp[-j\omega(2k-n)T]$$

This expression corresponds to the condition of loss-free (in informational sense) restoration of a continuous one-dimensional material flow signal from its discrete distribution.

In cases where the process model is characterized by a transfer function that has the form of a rational function with zeros available, it is necessary to divide the transfer function model to pole- and zero-containing parts, while for the pole-containing part (for the differential-free fragment of the structural scheme) the vector-matrix model of the multi-feeding system can be described both in a Frobenius form and with the help of a similarity transformation using actual physical signals recorded by measuring converters. The need for such a procedure arises in the analysis of feeding processes, when one or more feeders operate in a batch (discrete) mode, provided that the description of material flow signals is performed on the basis of Fourier models.

It should be noted that with modal controllers regulators supplied with virtual sensors (state observers), used in systems for which vector-matrix models are formed on the basis of augmented (i.e., original) structural schemes, calculations of the feedback matrix over the vector of state variables and the observer matrix are carried out according to corresponding Ackerman's formulas [2].

The data of current recording all of the material flow signals with synchronous visualization of them in the form of time-frequency distributions (time-frequency maps of controlled processes) make it possible to efficiently control and monitor – at the dispatch level – the processes occurring in different fragments of the mixture-producing aggregate, and the formation of an optimal control law with time-frequency distributions available (the latter registered in the wavelet medium and used instead of one-dimensional SV's), contributes to the improvement of transient processes quality in feeding subsystems, inside the equipment for transferring ingredient substances within the production space, and in the mixing unit - when readjusting their modes of operation.

As an example of the current registration (carried out synchronously with the control function) of material flows in the multi-feeding unit of the mixing aggregate, figure 1 shows the representation of the steady-state feeding process in the form of Wigner distributions at the outlet of the multi-feeding unit. The latter consists of one batch feeder, two spiral ones running continuously (current frequencies are of 0.20Hz, 3.23Hz, and 4.02Hz), and a spiral feeder that functions discretely in time with a central frequency of 6.9 Hz in the spectrum band and a frequency of 0.077 Hz as a dose-forming one.

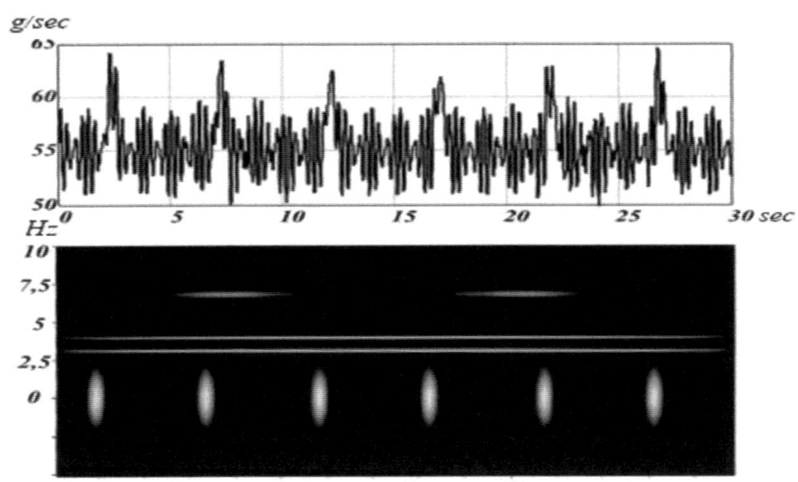

Figure 1. Display of the steady-state feeding process in the form of Wigner distributions at the outlet of the feeding unit.

Thus, realizing the procedure of synchronous feeding processes monitoring in the wavelet medium, combined with the controlling process, makes the automatic control system effective, and from the user's point of view – more informative regarding the semantics of the processes occurring in the mixture-producing unit.

5. Conclusion

To demonstrate the joint use of approaches based on the concepts of state space and wavelet transforms, a method for optimal control of a technological object in a state space medium is considered. In this case, the state variables are represented here in the form of time-frequency representations (wavelet maps).

The difference from the traditional method of synthesizing an optimal control system is that instead of one-dimensional vectors of state variables and of control in a closed system, multidimensional distributions of Cohen's class are used, in particular, time-frequency Wigner time-frequency representations.

The control plant is the feeding unit of the mixture-producing aggregate, and the optimal control system itself is strictly causal.

The computer-aided control system is formed on the basis of the similarity transformation method, which makes it possible to exclude the use of reduced state variable observers.

The Riccati equation is used to determine the state feedback matrix.

And finally, the results of monitoring the flow at the output of the multi-component feeding unit are presented. The example of this demonstrates the effectiveness of the computer-aided control system using multi-dimensional representations in the wavelet medium instead of *1D*- state variables.

Acknowledgements

All comments and suggestions by critical readers and editorial assistance are acknowledged with gratitude.

References

[1] Cohen L 1995 *Time-Frequency Analysis* (Englewood Cliffs: Prentice Hall) p 299
[2] Kim D P 2004 *Theory of Automatic Control* (Moscow: Fizmatlit) vol 2 p 464
Daubechies I 1992 *Ten Lectures on Wavelets: Regional Conference Series in Applied Mathematics* (Philadelphia PA: CBMS-NSF SIAM) p 357
[3] Fedosenkov B A, Nazimov A S and Sheboukov A V 2004 *Automation and Modern Technologies: Automation of Scientific Research and Production Processes* **8** 7–13
[4] Choi H I and Williams W J 1989 *IEEE Trans. Acoust., Speech, and Signal Proc.* **37(6)** 862–71
[5] Luenberger D G 2006 *Information Science* (Princeton NJ: Princeton University Press) p 448

IOP Conf. Series: Materials Science and Engineering **354** (2018) 012005 doi:10.1088/1757-899X/354/1/012005

Development of similarity theory for control systems

L P Myshlyaev[1], V F Evtushenko[1], K A Ivushkin[2] and G V Makarov[1]

[1]Siberian State Industrial University, 42 Kirova street, Novokuznetsk, 654007, Russia
[2]Consolidated Company "Sibshakhtostroy" LLC, 9 Kuznetskoye highway, Novokuznetsk, 654034, Russia

E-mail: mail@nicsu.ru

Abstract. The area of effective application of the traditional similarity theory and the need necessity of its development for systems are discussed. The main statements underlying the similarity theory of control systems are given. The conditions for the similarity of control systems and the need for similarity control control are formulated. Methods and algorithms for estimating and similarity control of control systems and the results of research of control systems based on their similarity are presented. The similarity control of systems includes the current evaluation of the degree of similarity of control systems and the development of actions controlling similarity, and the corresponding targeted change in the state of any element of control systems.

1. Introduction

Development and implementation of control systems are associated with the need for a wide application of modeling methods and, accordingly, models of objects and control systems. Adequacy and / or similarity indicators are used to evaluate the effectiveness of models, one of the important requirements for the model is to ensure the transfer of the results of model solutions to the original (full-scale control system) and its components.

According to the definitions of similarity and adequacy in [1, 2], the similarity is "a correspondence between the original and the model, under which the *rules of the transition* from model parameters to object parameters and vice versa are known, and the mathematical description (if it is known or can be obtained) can be transformed into an identical form". In its turn, the adequacy of the model is "its ability to display the specified properties and conditions of the object's operation with the required accuracy". From this it follows that the similarity of objects and control systems is a more general concept than adequacy, and includes *not only an assessment of the proximity of the model and the original, but also the procedures (methods, rules) of transition from model parameters to original parameters and vice versa.*

Particularly important is the requirement to transfer the solutions obtained in the model system to the full-scale one in cases when their coordinate and time dimensions do not coincide. In particular, when in the model system the controlled part is the physical model of the full-scale object. In such cases, the concept of adequacy of the model becomes insufficient for a well-founded transfer (recalculation) of model solutions to full-scale systems.

2. Features of the application of similarity theory for control systems

Recalculation of the results obtained from one system to another in theory and practice of control occurs quite often in the following classes of problems:

Content from this work may be used under the terms of the Creative Commons Attribution 3.0 licence. Any further distribution of this work must maintain attribution to the author(s) and the title of the work, journal citation and DOI.
Published under licence by IOP Publishing Ltd

- research of processes (technical, socio-economic, etc.) on different types of models for the creation of new processes and aggregates, productions, organizational structures [3];
- directly in control in classes of structures of control systems with predictive models, for example, [4];
- in field tests and adjustment of control systems, in particular [5].

Thus, based on the definitions given in [1, 2], we can conclude that a *reasonable transfer (recalculation) of results from one control system to another is possible only if they are similar*. To determine the similarity of the design characteristics of uncontrolled objects and the processes of energy and substance conversion that take place in them, the methods and criteria of the traditional similarity theory are now well developed and widely used [1]. However, they can not be successfully used to determine the similarity of control systems [6], which *requires significant improvements and development, not only for the systems as a whole, but also for their main components* [6, 7].

This conclusion is also confirmed by the fundamental difference in the structure of mathematical models used in the traditional similarity theory and control theory. The modern theory of similarity relies on mathematical models of fundamental disciplines [1]. But in the theory and practice of automatic control systems development, a special class of models is used that differs significantly "from models used in such fundamental disciplines as physics, chemistry" [8] and reflecting "the most important properties of the control object with respect to predefined input and output variables" [8]. In other words, they should jointly display both the dynamics of the channels for transforming external actions, and the properties of the actions themselves [9].

At the same time, it is possible to note particular examples of the application of certain similarity indices in solving engineering problems of analysis and synthesis of control systems. First, it is the ratio of the time of pure lag to the time constant of inertia, used, in particular, in engineering methods of selecting and tuning regulatory laws [10]. Second, when estimating the effective work areas of control systems with feedback, the time of the net delay of the dynamic channel for converting the control actions is compared with the interval of the autocorrelation function decrease of the perturbation object brought to the output [10].

3. Basic ideas of the similarity theory of control systems

Studies related to the development of similarity theory with respect to control systems should target, in our opinion,

- development of the basic concepts and axioms (in particular, in the form of statements) of similarity of control systems reflecting the required properties of not only such control systems as a whole, but also its individual components, including external influences;

- the development of procedures, methods and algorithms for estimating and controlling the similarity of control systems, which, of course, allow the research and adjustment problems of one such system (for example, model) to be solved, and then the results to another to be recalculated;

- application of the developed procedures, methods and algorithms for full-scale (technical, socio-economic and other) systems to effectively solve specific control tasks.

The following statements are the basis for the development of similarity theory of control systems.

Statement 1. Control systems are similar if the performance indicators (criteria) of their operation are equal to the specified value.

This statement for the j-th and l-th control systems can be written in the form

$$\left\| q_j^n\{(t-T),t\} - q_l^n\{(t-T),t\} \right\| \le \delta q_{jl}^{n*};\ j \ne l;\ j=\overline{1,J};\quad l=\overline{1,L}, \tag{1}$$

where $\| \ \|$ is the measure of proximity; $q_j^n\{(t-T),t\}$ and $q_l^n\{(t-T),t\}$ – the values of the normalized target indicators of functioning at a given time interval T, respectively, of the j-th and l-th control systems; t – continuous or discrete time; δq_{jl}^{n*} – the value determined by the maximum permissible (specified) threshold of proximity of the normalized indicators $q_j^n\{(t-T),t\}$ and $q_l^n\{(t-T),t\}$, for which the effectiveness of these systems is assumed to be the same; J and L – the number of control systems.

Rationing of target indicators is carried out, for example, by the expression

$$q^n\{(t-T),t\}=\frac{q\{(t-T),t\}}{q_{max}}, \tag{2}$$

where q_{max} – the maximum possible value of the target indicators of control systems.

Statement 2. *The degree of similarity of control systems is a quantitative measure of proximity of performance indicators of these systems control.*

An example of the method of quantitative estimation of the similarity degree is the method of correlated processes described in [11], where it is proposed to "establish the degree of correspondence between the initial and simplified systems ... by the correlation moments of statistical values of the probability characteristics" of these systems. In addition, the degree of similarity can be estimated on the basis of relation (1).

Statement 3. The objects in control systems are similar if they satisfy the conditions for a joint similarity of controlled and uncontrolled external effects and dynamic channels to their transformation. This condition is written in the form of equality with a specified small number of effects of changes in the input actions on the output actions of the control object.

Uncontrolled effects can be estimated by calculation or experimentally by bringing them to the output [10] or to one of the control inputs [12] of the control object.

The condition of similarity is written, for example, using the following equations

$$\left\|o_j^n\{(t-T),t\}-o_l^n\{(t-T),t\}\right\|\le \delta o^*;\ j\ne l;\ j=\overline{1,J};\quad l=\overline{1,L}; \tag{3}$$

$$o_j\{(t-T),t\}=\int_{t-T}^{t}\varphi_{ej}(\theta)\{i_j(t-\theta)\}d\theta; \tag{4}$$

$$o_l\{(t-T),t\}=\int_{t-T}^{t}\varphi_{il}(\theta)\{i_l(t-\theta)\}d\theta; \tag{5}$$

$$i(t)\in\{e(t),c\ (t)\}, \tag{6}$$

where $i(t)$, $e(t)$, $c(t)$, $o(t)$ – are respectively the input, external, control and output actions of the control object at the moment of time t; T is the interval for estimating the response of the control object to the change in the input actions; $\varphi_i\{\cdot\}$ – the dynamic operator of the conversion channel of input into output actions; δo^* – the value determined by the maximum permissible threshold of proximity of normalized values $o_j^n\{(t-T),t\}$ and $o_l^n\{(t-T),t\}$; θ is an auxiliary variable; J and L are the number of control systems.

If the conditions for the similarity of control systems, for example, according to the relation (1), are not satisfied, then the degree of similarity of the control systems under consideration is calculated. Quantitatively, it can be estimated either by a measure of the difference δq_{jl}^n obtained by the relation (1) in dimensionless form, or by the correlation coefficient between the target indicators reflecting the efficiency of these systems functioning [11]. If the value of the similarity degree of the j-th and the l-th control systems does not correspond to the required one, then the possibility of achieving similarity of control systems due to the subsequent target change in the properties of the *j*-th and *l*-th control systems and their actions is estimated. The process of action on the control systems and their individual components aimed at fulfilling the similarity conditions will be called the *similarity control of systems and their components.*

Similarity control of control systems is based on the following statements.

Statement 4. Control systems are potentially similar if the exact or approximate equality of the values of their target indicators (criteria) of control efficiency can be achieved due to the targeted impacts on the change in the dynamic properties of systems as a whole and (or) their elements, including external actions.

Statement 5. Control objects are potentially similar if the condition of joint similarity of their objects and input actions on a finite interval of time T can be achieved due to targeted actions on the change in the dynamic properties of controlled and uncontrolled external actions and channels for their transformation.

Statement 6. Control systems are fundamentally not similar if the exact or approximate equality of the values of their target indicators (criteria) of control efficiency can not be achieved due to targeted actions on the change in the dynamic properties of systems as a whole (without changing their structure) and / or their elements, including external actions.

4. The results of development of the similarity theory of control systems

The statements formulated are the basis for the development of methods and algorithms for evaluating and controlling the similarity of control systems and, accordingly, for the research of control systems based on their similarity.

It is assumed that all the structural diversity of control systems can be represented in terms of their similarity in the form of three classes: *similar, potentially similar and fundamentally different control systems.*

The process of control of systems similarity includes the current estimation of the similarity degree of control systems and, if similarity is violated, the development of control actions by the similarity that purposefully changes the state of any element of control systems. It should be noted that such purposeful change in the properties of one of the control systems can significantly expand the field of effective application of model control systems in solving practical problems, in particular, related to the study and adjustment of full-scale systems.

5. Conclusions

1. Transfer of the results of development and research (adjustment) from one control system to another is justified only if they are similar.

2. Statements based on the equality or proximity of the values of performance indicators of control systems are expedient to use for the creation of procedures, methods and algorithms of estimating and controlling similarity.

3. Similarity control of systems makes it possible to ensure the fulfillment of similarity conditions for potentially similar control systems and, thus, to expand the area of effective application of methods and algorithms for their development and research.

References

[1] Venikov V A 1976 *Theory of Similarity and Modeling* (Moscow: Higher School)

[2] Venikov V A 1987 *Similarity and Modeling. Basic Concepts. Terminology. Committee of Scientific and Technical Terminology* (Moscow: Nauka) issue 113

[3] Mochalov S P, Ivushkin A A, Myshlyaev L P, Wenger K G et al 2012 *Coal* **10 (1039)** 49–53

[4] Krasovskii A A 1979 *Automation and Telemechanics* **2** 156–162

[5] Evtushenko V F, Myshlyaev L P, Makarov G.V et al 2016 *IOP Conference Series: Earth and Environmental Science* **45** 012010

[6] Myshlyaev L P, Evtushenko V F et al 2010 *Izv. Vvuzov. Ferrous Metallurgy* **12** 65–67

IOP Conf. Series: Materials Science and Engineering **354** (2018) 012005 doi:10.1088/1757-899X/354/1/012005

[7] Myshlyaev L P, Evtushenko V F et al 2012 *Izv. Vuzov. Ferrous Metallurgy* **12** 58–60

[8] Emelyanov S V and Korovin S K 1997 *New Types of Feedback: Control under Uncertainty* (Moscow: Nauka, Fizmatlit)

[9] Myshlyaev L P, Fayrushin Sh A et al 2016 *Proc. Int. Conf. in Knowledge based Tech. for the Development and Use of Resources* (Novokuznetsk: SibSIU) pp 284–286

[10] Rotach V Ya 1973 *Calculation of the Dynamics of Industrial Automatic Control Systems* (Moscow: Energy)

[11] Pugachev V N 1973 *Combined Methods for Determining Probabilistic Characteristics* (Moscow: Soviet Radio)

[12] Barkovskiy V V, Zakharov V N and Shatalov A S 1969 *Methods of Synthesis of Control Systems* (Moscow: Mashinostroyeniye)

IOP Conf. Series: Materials Science and Engineering **354** (2018) 012006 doi:10.1088/1757-899X/354/1/012006

Improving reliability of aggregation, numerical simulation and analysis of complex systems by empirical data

Boris S Dobronets and Olga A Popova

Institute of Space and Information Technology, Siberian Federal University, Kirenskogo 26, Krasnoyarsk, 660074, Russia

E-mail: address BDobronets@yandex.ru

Abstract. The paper considers a new approach of regression modeling that uses aggregated data presented in the form of density functions. Approaches to Improving the reliability of aggregation of empirical data are considered: improving accuracy and estimating errors. We discuss the procedures of data aggregation as a preprocessing stage for subsequent to regression modeling. An important feature of study is demonstration of the way how represent the aggregated data. It is proposed to use piecewise polynomial models, including spline aggregate functions. We show that the proposed approach to data aggregation can be interpreted as the frequency distribution. To study its properties density function concept is used.Various types of mathematical models of data aggregation are discussed. For the construction of regression models, it is proposed to use data representation procedures based on piecewise polynomial models. New approaches to modeling functional dependencies based on spline aggregations are proposed.

1. Introduction

The linear regression analysis is often used to discovery dependencies of the empirical data. Properties of empirical data including the type and level of uncertainty significantly affect on the results of the simulation. It is well-known, that the random uncertainty concept addressed by probability theory plays a fundamental role to study data uncertainty [1, 2].

Aggregation is quite a popular method of converting big data [3-6]. For example, the application of the histogram allows to reduce dimension of the data set and level of uncertainty and to significantly increase the efficiency of numerical calculations. It is important to note that the histograms are examples of the symbolic data using in the Symbolic data analysis [1, 4, 7].

Symbolic Data Analysis and Data Mining use the histograms to study a variety of different processes and are applied to modeling the variability of quantitative characteristics.

Histogram data models and histogram regression models based on the Symbolic analysis is a new important direction to discover knowledge in a data base. Billard L., Diday E. proposed the symbolic data type named histogram-valued variables to employ for regression modeling [1, 7].

This problem becomes more complicated if large amounts of data are processed. In this case it is useful to look at the empirical data in an aggregated form. Aggregation is a popular method of converting data. For example, the application of the histogram allows to reduce dimension of data set and level of uncertainty and to significantly increase the efficiency of numerical calculations. It is important to note that histograms are the examples of symbolic data used in the Symbolic data analysis

Content from this work may be used under the terms of the Creative Commons Attribution 3.0 licence. Any further distribution of this work must maintain attribution to the author(s) and the title of the work, journal citation and DOI.
Published under licence by IOP Publishing Ltd

[7]. Symbolic Data Analysis and Data Mining use histograms to study a variety of different processes and are applied for modeling the variability of quantitative characteristics. Histogram data models and histogram regression models based on the Symbolic analysis is a new important direction to discover knowledge in a data base.

In our study we consider a new approach to regression modeling using input data aggregation. To develop our approach for performing efficient aggregation we employ piecewise polynomial aggregation functions, including piecewise linear functions and piecewise constant functions. Histogram is a good example of piecewise constant functions of which are perfectly employed in our study.

To examine the structure of data aggregation we use the probability density functions (PDF). The concept of the mathematical aggregation functions is used to the regression modeling. To illustrate this we will regard the spline aggregation function in more detail. This approach will allow to employ the density function models as input and output data. It is of further importance that, the data uncertainty is studied to identify the relationship between the input and output characteristics when the input probability density functions are unknown. Thus, in order to describe any specific PDF we need to consider their spline interpolation.

In this work we propose a new linear regression model named a PDF-valued variable regression. The abbreviated form of the regression model is called a Distributions Regression. If we use a spline aggregation model it is named a PDF-spline valued variable Regression Model and a Distributions Regression in shot. The following statements confirm the justification of PDF-spline models. The application of the spline procedures allows big data aggregating reduce the level of uncertainty and to significantly increase the efficiency of numerical calculations. These splines allow considerably accurate representions of the arbitrary distribution.

To demonstrate the degree of the relevance of the proposed methods to reality, we developed a theoretical study and provided numerical examples to illustrate it. With this we propose a conclusive discussion of this approach applicability to the uncertainty treatment and big data processing. The comparison of NPA and Monte Carlo method showed good agreement of the results. At the same time, numerical experiments demonstrate that the PDF arithmetic is more than hundred times faster than the Monte Carlo method [8-10]. As a result, the NPA approach can be successfully applied for solving computational and engineering problems [9,10,11].

2. Reliable wstimates of the accuracy of aggregation

In those cases where data can be interpreted as frequency distributions, it is advisable to use methods for constructing the probability density function for aggregation. It is remarkable that the histogram stood as the only nonparametric density estimator until the 1950s, when substantial and simultaneous progress was made in density estimation and in spectral density estimation. During the following decade, several general algorithms and alternative theoretical modes of analysis were introduced by Rosenblatt, Parzen, and Cencov [12].

Next, consider using Richardson's extrapolation to improve the accuracy of the kernel estimator [13].

The basic kernel estimator may be written compactly as [12]

$$\hat{f}_h(x) = \frac{1}{Nh}\sum_{i=1}^{N} K\left(\frac{x-\xi_i}{h}\right) = \frac{1}{Nh}\sum_{i=1}^{N} K_h(x-\xi_i),$$

where $K_h(t) = K(t/h)/h$.

Note

$$K_h(x,\xi_i) = K\left(\frac{x-\xi_i}{h}\right).$$

IOP Conf. Series: Materials Science and Engineering **354** (2018) 012006 doi:10.1088/1757-899X/354/1/012006

where ξ is a random variable with probability density function $f(x)$.
Then expected value

$$\mathrm{E}[\hat{f}(x)] = \mathrm{E}[K_h(x,\xi)]$$

and variability

$$\sigma_N = \mathrm{Var}[\hat{f}(x)] = \frac{1}{N}\mathrm{Var}[K_h(x,\xi)].$$

Suppose that the kernel K satisfies the requirements

$$\int_{-\infty}^{\infty} K(\eta)d\eta = 1, \int_{-\infty}^{\infty} \eta K(\eta)d\eta = 0$$

and

$$\int_{-\infty}^{\infty} \eta^3 K(\eta)d\eta = 0.$$

Denote

$$\int_{-\infty}^{\infty} \eta^2 K(\eta)d\eta = \sigma^2.$$

Define $f^h(x)$ as follows

$$f^h(x) = \mathrm{E}[\hat{f}_h(x)] = f(x) + \sigma^2 h^2 f''(x)/2 + O(h^4). \quad (1)$$

and $f^{2h}(x)$ as

$$f^{2h}(x) = \mathrm{E}[\hat{f}_{2h}(x)] = f(x) + 4\sigma^2 h^2 f''(x)/2 + O(h^4). \quad (2)$$

Let we apply the Richardson's extrapolation to $f^h(x)$ and $f^{2h}(x)$ [7]. In the next stage, we multiply (1) on 1/4 to subtract the result from (2). Excluding $\sigma^2 h^2 f''(x)/2$ from (4) and (5), we get

$$f(x) = \frac{4}{3}f^h(x) - \frac{1}{3}f^{2h}(x) + O(h^4).$$

Noting that we have constructed the approximation to the function $f(x)$

$$f_{cor}^h(x) = \frac{4}{3}\hat{f}^h(x) - \frac{1}{3}\hat{f}^{2h}(x). \quad (3)$$

with the accuracy $O(h^4)$.

In figure 1 we represent numerical example. The solid line is exact probability density function *f(x)*, a – kernel estimator of probability density function, b – correction of kernel estimator function by Richardson's extrapolation.

(a) (b)

Figure 1. Improving the accuracy of the probability density function estimation.

Thus, successively setting $z \in \omega$, we obtain the values $f^h_{cor}(x_i) = f(x_i) + O(h^4)$. Further, using the obtained values, we can construct systems of linear algebraic equations for constructing a cubic spline [14].

On the other hand, applying the Runge rule, we can obtain the estimate [11]

$$f''(x) = 2(f^h(x) - f^{2h}(x))/(3\sigma h^2) + O(h^2)$$

or

$$\| \hat{f}'' \| = \frac{2}{3\sigma h^2} \| \hat{f}^h - \hat{f}^{2h} \| \tag{4}$$

Thus, a posteriori estimate is constructed for the second derivative of the density function. This allows one to obtain an estimate for the accuracy of the approximations constructed.

3. Spline aggregation

A spline is a sufficiently smooth polynomial function that is piecewise-defined, and possesses a high degree of smoothness at the places where the polynomial pieces connect (which are known as knots). We will consider the probability density of the random variables as an approximated spline.

Prior to consideration of spline aggregation, we propose study of mathematical models applicable for the representation of splines and will discuss the interpolation questions with their application.

Let $\omega = \{x_0 < x_1 < x_2 < \ldots < x_n\}$ be mesh and interpolation conditions

$s(x_i) = f(x_i), i = 0, \ldots, n.$

Where boundary conditions are given as

$s'(x_0) = 0, s'(x_n) = 0.$

The cubic spline on a mesh $\{x_i\}$ with step $\bar{h} = \max(x_{i+1} - x_i), i = 0, \ldots, n$ satisfies the estimate

$$\| f^{\nu} - s^{\nu} \| \le \bar{h}^{4-\nu} \| f^{(4)} \|, \nu = 0,1,2. \tag{5}$$

The task of the spline construction is reduced for solving a system of linear algebraic equations with a tridiagonal matrix [14]

$$\begin{aligned}
&\lambda_j m_{j-1} + 2m_j + \mu_j m_{j+1} = d_j, \\
&2m_0 + m_1 = 3(f_1 - f_0)/h_1 - h_1 z_0^2/2, \\
&2m_n + m_{n-1} = 3(f_n - f_{n-1})/h_n + h_n z_n^2/2, \\
&d_j = 3\lambda_j (f_j - f_{j-1})/h_j + 3\mu_j (f_{j+1} - f_j)/h_{j+1}, \quad j = 1, \ldots, N-1.
\end{aligned} \tag{6}$$

where $m_i = s'(x_i)$. As a result, a cubic spline on the intervals $[x_{j-1}, x_j], j = 1, \ldots, N$ will have the following representation [8]:

$$s(x) = m_{j-1}(x_j - x)^2 (x - x_{j-1})/h_j^2 - m_j (x - x_{j-1})^2 (x_j - x)/h_j^2 +$$

$$+f_{j-1}(x_j-x)^2(2(x-x_{j-1})+h_j)/h_j^3++f_j(x-x_{j-1})^2(2(x_j-x)+h_j)/h_j^3,$$

Let consider the spline approach to build regression model with the Distributions-valued variables. This approach is useful due to the following reasons. Underlying of this approach is the notion of the spline. The spline can be regarded as a mathematical object that is easy to describe and calculate the mathematical procedures and operations, in the process of maintaining the essence of data frequency distribution.

Since the spline is a piecewise polynomial function then it can be regarded as a data aggregation function in aggregation issues. Aggregation function performs numerical calculations on a data set and returns the spline values. Splines are useful for data uncertainty analysis due to fact that they adequately represent the random distribution of random variables.

Despite its simplicity, the spline also covers all possible ranges of probability density function estimation. Simple and flexible spline structure greatly simplifies their use in numerical calculations and it has a clear visual image, which is useful for analytical conclusions. It is important to note that the construction of regression models with aggregated inputs require the use of appropriate numerical procedures. To this end, we consider numerical probabilistic analysis. We propose to use of the numerical probabilistic analysis to compute the arithmetic operations for the aggregated data and to apply for regression modeling.

Consider the problem of aggregating data by splines. For this purpose, we construct a spline s approximating the density function f, so that the estimate

$$\| f-s \| \leq Ch^4.$$

Thus, successively assuming $z \in \omega$, we obtain the values $f_{cor}^h(x_i)=f(x_i)+O(h^4)$ and the system of linear algebraic equations (8) for constructing a cubic spline. To improve the reconstruction accuracy of probability density at the point z we use the combination of kernel assessments with the parameters h and $2h$.

As an evidence we refer to Schweizer who states, that "distributions are the numbers of the future" [15]. Thus, instead of simplifying them, it seems better to propose methods which deal with distributions directly. In order to do this, one has to determine how to represent the observed distributions.

In our study we propose to represent them by using a piecewise polynomial aggregation function, as long as it offers a good tradeoff between simplicity and accuracy.

4. Distributions Regression

Consider a linear model

$$Y=a_0+\sum_{i=1}^{n} a_i X_i+\varepsilon,$$

where X_i, $i=1,\dots,n$ are independent predictor variables, Y is a dependent variable, ε is an error. From the observed values of Y_j $X_{i,j}$ after the aggregation of the density Y, X_i are represented by splines: S_y, S_i.

We shall seek the unknown parameters a_i, $i=0,1,\dots,n$ starting from the minimum of the functional

$$\Phi(a_0,a_1,...,a_n)=\| S_y-(a_0+\sum_{i=1}^{n}a_iX_i)\|_2\to \min.$$

By virtue of the independence of X_i, numerical operations on density functions can be used to calculate the functional $\Phi(a_0,a_1,...,a_n)$. The minimization of the functional $\Phi(a_0,a_1,...,a_n)$ can be carried out by the method of steepest descent.

Let us consider model problem n=2

$$Y=a_0+\sum_{i=1}^{n}a_iX_i+\varepsilon,$$

For numerical realization, X_1, X_2 were generated as sums of random variables with an Irvine-Hall distribution n = 3 and shifted by 1 and 2, respectively, ε with probability density function $(|2x|-1)^2(2|2x|+1)$ with support $[-0.5,0.5]$.

The variable Y was constructed as follows $Y=X_1+X_2+\varepsilon$.

The minimization of the functional $\Phi(a_0,a_1,a_2)$ was carried out by the method of steepest descent. For $a_0=-0.089$, $a_1=1.031$, $a_2=1.029$, the value $\Phi(a_0,a_1,a_2)$ did not exceed the value $0.3\cdot 10^{-3}$.

Thus, a numerical example showed the possibility of using distributions regression.

5. Conclusion

Although there are many ways of data aggregation, including simple average, we argue that the use of piecewise linear and piecewise polynomial aggregation functions will offer a more informative representation of the variability in the data, than other forms of data aggregation. To prove their thesis, we considered the aggregation procedure based on the histogram time series. Using these types of data aggregation for preprocessing and regression modeling you contribute to the reliability of the study of natural systems and processes. The spatial and time aggregation procedures help to reduce the amount of computation in data processing and are an important basis for the extraction of useful knowledge from large volumes of data. Developed methods reduce the level of uncertainty in the information flow; significantly reduce the processing time and the implementation of numerical procedures. This approach allows to the mode of interactive visual modeling to provide the necessary data for operational decision making under remote surveillance techniques and distributed object systems. In concluding the discussing about the applicability of this approach to practice we say about the advantage for uncertainty treatment and big data processing. Using the proposed model, applications with real and simulated data are presented.

References

[1] Dias S and Brito P 2015 *Linear Regression Model with Histogram-Valued Variables.* (wileyonlinelibrary.com). DOI:10.1002/sam.11260
[2] Koenker R 2005 *Quantile regression. Cambridge university press*
[3] Färe R, Grosskopf S and Primont D 2007 *Aggregation, Efficiency, and Measurement* (Springer, New York)
[4] Arroyoa J and Maté C 2009 *International Journal of Forecasting* 192–207
[5] Nava J 2012 *Journal of Uncertain Systems* vol 6 **2** 84–85
[6] Tchangani A 2013 *Journal of Uncertain Systems* **7(2)** 138–151

IOP Conf. Series: Materials Science and Engineering **354** (2018) 012006 doi:10.1088/1757-899X/354/1/012006

[7] Billard L and Diday E 2006 *Symbolic Data Analysis: Conceptual Statistics and Data Mining.* (Wiley)

[8] Dobronets B S, Krantsevich A M and Krantsevich N M 2013 *J. Sib. Fed. Univ. Mathematics & Physics* **6(2)** 168–173

[9] Dobronets B and Popova O 2016 *Numerical Probabilistic Approach for Optimization Problems. Scientific Computing, Computer Arithmetic, and Validated Numerics. Lecture Notes in Computer Science* 9553 (Springer International Publishing) pp 43–53

[10] Dobronets B S and Popova O A 2014 *Reliable Computing* vol 19 274–289

[11] Dobronets B S and Popova O A *Journal of Siberian Federal University – Engineering & Technologies* **9**(7) 960-971

[12] Scott R W 2015 *Multivariate Density Estimation: Theory, Practice, and Visualization* (New York: John Wiley & Sons)

[13] Dobronets B S and Popova O A 2017 *Journal of Siberian Federal University – Mathematics & Physics* **10** (1) 16–21

[14] Ahlberg J H, Nilson E N and Walsh J L 1967 *The theory of Splines and Their Applications* (New York: Academic Press)

[15] Schweizer B 1984 *Distributions Are the Numbers of the Future, in Proceedings of The Mathematics of Fuzzy Systems Meeting* ed A di Nola and A Ventre, (Italy: University of Naples) pp 137–149.

IOP Conf. Series: Materials Science and Engineering **354** (2018) 012007 doi:10.1088/1757-899X/354/1/012007

About the algorithm for construction of coordinated university timetables

A S Dobrynin, S M Kulakov and E N Taraborina

Siberian State Industrial University, 42 Kirova street, Novokuznetsk, 654007, Russia

E-mail: kulakov-ais@mail.ru

Abstract. The factual description of the task and an algorithm for drawing up a coordinated timetable of academic work of the faculty and students at the level of department (local timetable) is presented, as well as the procedure for integrating private schedules, i.e. the formation of a university-wide timetable. Coordination of the latter has not only spatio-temporal in nature, but also takes into account the preferences (interests) of agents (users and performers of works).

1. Introduction

A complete mathematical model of the elements of academic work and university timetable, is not given here due to its cumbersome nature. It can be found in [1]. The most important elements of the department timetable are:

a) the disciplines taught at the department $\overline{DS^d}$;

b) faculty of the department $\overline{F^d}$;

d) groups of students $\overline{G^d}$ taught by the faculty;

e) classrooms $\overline{Cl}$, laboratories $\overline{La}$ and other resources assigned to the department.

Preferences of the performers (teachers and groups of students) can be formally represented by the corresponding matrices $Mr[d,t]$, $Mr[d,t,(cl\,|\,la)]$ where d – the day, t – the time slot (or "class"), cl – classroom, la – laboratory.

The essence of the problem is formed as follows. Agents of the department – faculty and groups of students; resources of the department – laboratories (la), classrooms (cl) and other collective workplaces; types of academic work – lectures, practical classes, laboratory classes, etc., are given. The requirements for specific performers of works (teachers and groups of students) are defined. The department's assets are linked by relationships, which are well illustrated by a typical academic workload form, see below (figure 1).

Drawing up of curricula and preparation of teachers academic workloads is a number of related tasks that are indirectly present in the task of constructing the academic work timetable. In order to avoid collisions when merging local (departmental) timetables, it is desirable that the elements $\overline{DS^d}, \overline{F^d}, \overline{G^d}, \overline{Cl^d}$ from different departments do not overlap with each other.

Academic secretaries of different departments who form local timetables and academic workload should mainly use only their own assets when planning educational works and the range of timeslots allocated to them. Such requirements can be met in practice. For example, the prerequisites for a simple merging of departmental timetables into the university-wide one without collisions can be

Content from this work may be used under the terms of the Creative Commons Attribution 3.0 licence. Any further distribution of this work must maintain attribution to the author(s) and the title of the work, journal citation and DOI.
Published under licence by IOP Publishing Ltd

formulated in a semantic statement: "if the academic secretary of a graduate department N is using only its own assets {groups, faculty, subjects and classrooms} and the timeslots allocated by the administration, then there will be no collisions when merging the local departmental timetables into the university timetable".

No.	Subject name	Group No.	Number of students	Autumn semester													
				Lectures	Practical classes	Consultations	Laboratory classes	Credits	Exams	Graduation project	Term project	Internship	State Examination Board	Classes with postgraduates	Attending classes	Control of students independent work	Supervision of R&D
1.	Programming	ISU-14	16	18	20			8									
2.	Operating system	ISU-12	17	36			18	6									
3.	Information security	IAT-14	20	16	18				6								
4.	Programming	IAT-13	18	18			18		5								

Aggregate work, work $_{aggr}$
work$_{aggr}$ = { GUID ID – discipline identifier;
GUID ID – teacher identifier;
GUID ID1...GUID ID5 – student group identifier;
pCnt – number of points on the time grid (two weeks);
szCnt – the amount of work to link audiences;
jType – type of work (lectures, pract. classes, labs, etc.);
isAggregate=true; }

Atomic work, work $_{atm}$
work$_{atm}$ = { GUID ID – discipline identifier;
GUID ID – teacher identifier;
GUID ID1...GUID ID5 – one group, other - {null};
pCnt – number of points on the time grid;
szCnt – the amount of work to link audiences;
jType – type of work (lectures, pract. classes, labs, etc.);
isAggregate=false; }

Figure 1. Structure of the teacher's academic workload.

1) On the basis of faculty academic workload of the department to form academic works (indicating for each of them the following information: subject, teacher, location, desired time) with certain properties that can be distributed on the time grid. The work must correspond to one type of activities (lecture, practical class, laboratory work, etc.), as well as one or more groups of students.

2) To form a cluster of restrictions, to which the impossibility of using a certain asset (teacher, classroom, student group, subject) several times, within the same time slot ("classes") [d,t] belongs to.

3) To distribute the works on the time grid so that all work time counters (or time-slot counters) would turn into zero values, and often contradictory requirements of the performers would be taken into account with regard to "hard and soft" restrictions on the final timetable.

Despite the "simple" factual description, the formulated problem is complex, both from the position of algorithmic machine realization, and from the point of view of computational growth (in other words, it is NP-difficult). However, the relatively low dimensionality for a separate department allows us to obtain a coordinated solution.

2. Algorithm for priority distribution of training activities

The authors propose a general algorithm (figure 2) for the priority distribution of academic activities, which can be used to synthesize timetable agreed with performers of timetables. In fact, a human-machine procedure is proposed, which consists of three main stages:

1) Formation and actualization of information for the preparation of the classes timetable, including a list of all departmental assets and a list of requirements (wishes) of performers (agents);

2) Machine synthesis of the departmental timetable, according to the criteria of minimizing the "gaps" and maximizing the requirements of agents.

IOP Conf. Series: Materials Science and Engineering **354** (2018) 012007 doi:10.1088/1757-899X/354/1/012007

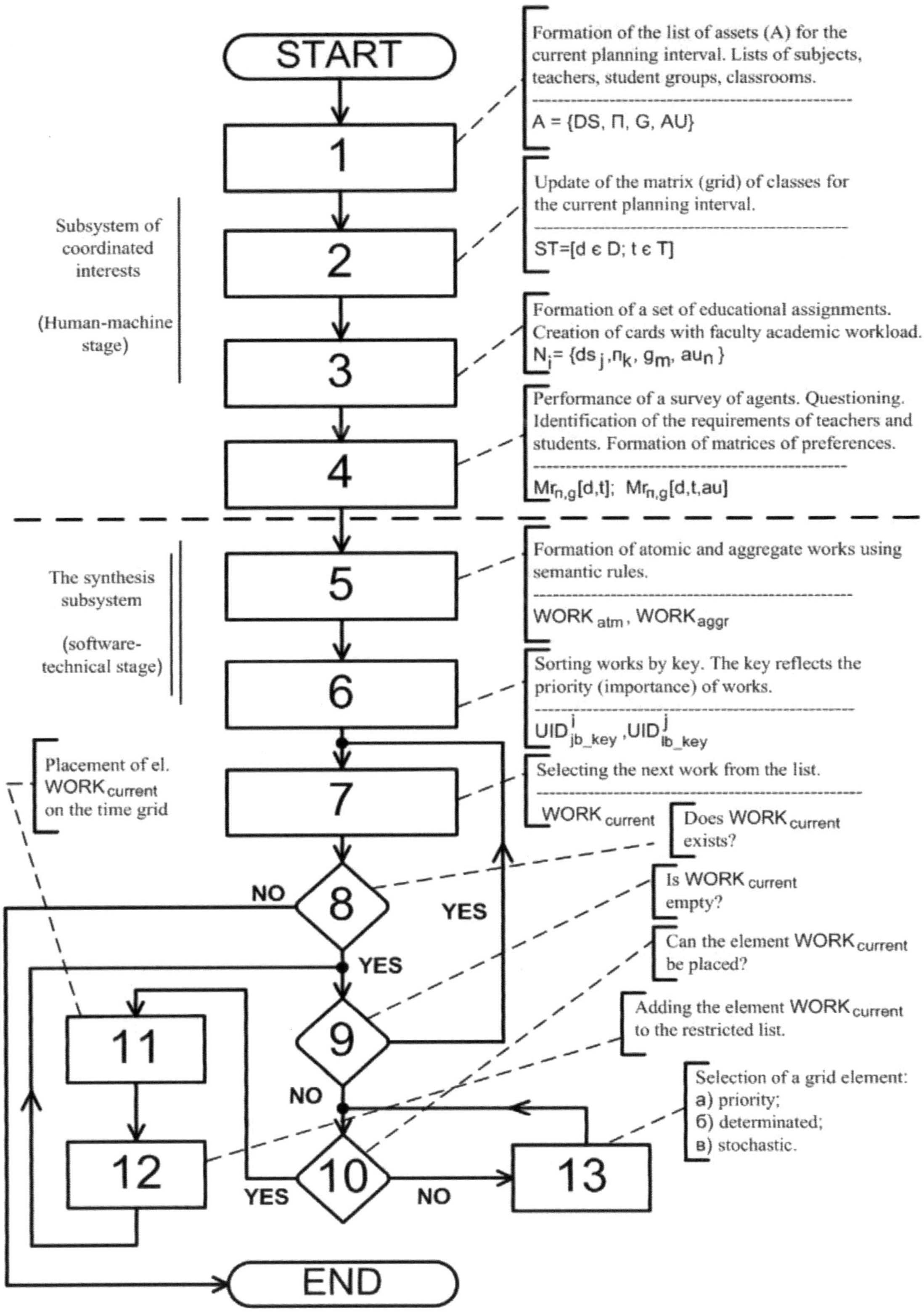

Figure 2. Consolidated procedure for the priority distribution of academic activities.

3) The merging of departmental timetables into a university one is carried out in the absence of collisions of previously formed fragments. If it is impossible to meet important requirements for assets, it is advisable to allocate additional timeslots in the timetable.

One of the most important requirements for the algorithm is the formation of correct, qualitative timetables at each stage of synthesis. This need is caused by a large number of information links, varying preferences of performers and indirect dependencies. The keys of priorities sorting are defined by the expressions (1) (2):

$$UID_{jb\_key} = \sum_{d=1}^{D} \sum_{s=1}^{S} Mr[d,s] \tag{1}$$

where UID_{jb_key} – the key of sorting for the greedy algorithm in the "teacher –group – academic load" space.

$$UID_{lb\_key} = Mr[d,s] \tag{2}$$

3. Conclusion

The article considers the task of synthesizing coordinated departmental timetables [1, 4], taking into account the wishes (requirements) of the performers. Consolidated human-machine procedure can be implemented as a part of large algorithmic complex of timetable construction [2] or within a separate software product. Modifications and processing of individual elements of the algorithm are allowed as it is implemented in software, as well as in the event of technical problems or organizational problems.

References

[1] Dobrynin A S et al 2013 *Scientific Review* **9** 95–101
[2] Dobrynin A S et al 2013 *Scientific Review* **8** 93–101
[3] Burkov V N, Korgin N A and Novikov D A 2009 *Introduction to the Theory of Control of Organizational Systems* (Moscow: Librocom) p 264
[4] Koinov R S et al 2013 *Bulletin of the Development of Science and Education* **6** 23–27

IOP Conf. Series: Materials Science and Engineering **354** (2018) 012008 doi:10.1088/1757-899X/354/1/012008

Problems in modernization of automation systems at coal preparation plants

L P Myshlyaev[1], M V Lyakhovets[2], K G Venger[3], I A Leontiev[3], G V Makarov[2] and A S Salamatin[2]

[1]"Scientific-research Center for Control Systems" LLC, 42 Kirova street, Novokuznetsk, 654007, Russia
[2]Siberian State Industrial University, 42 Kirova street, Novokuznetsk, 654007, Russia
[3]"Stroyservis" CJSC, 121 Kuznetsky ave., Kemerovo, 650055, Russia

E-mail: mail@nicsu.ru

Abstract. The factors influencing the process of modernization (reconstruction) of the automation systems at coal preparation plants are described. Problems such as heterogeneity of existing and developed systems, planning of reconstruction of a technological complex without taking into account modernization of automated systems, commissioning without stopping the existing technological complex, as well as problems of conducting procurement procedures are discussed. The option of stage-by-stage start-up and adjustment works in the conditions of modernization of systems without long stops of the process equipment is offered.

1. Introduction

Currently, only in Kemerovo region operates more than 40 preparation plants of different production capacity. Some of them have a long history – they were launched more than half a century ago. The requirement of owners to constantly increase the production capacity, as well as the rapid development of modern enrichment technologies, lead to the need for reconstruction of technological complexes of preparation plants (PP). Usually the issue of reconstruction or renovation and automated control systems is solved during the reconstruction of technological complexes (TC).

2. The main problems of modernization

When modernization of automation systems for coal preparation plants is carried out, it is necessary to take into account a number of important factors that are encountered in practice. Ignoring these factors leads to an increase in costs and timing of modernization (figure 1).

Heterogeneous systems. During design of a new PP and, in the future, development of modernization projects, often, when ordering large technological equipment (filter presses, jigging machines, flotation machines, etc.), no requirements for equipment suppliers in the part of local automation systems are specified (usually questionnaires of manufacturers and suppliers do not contain relevant information at all). In this connection, deliveries are made without taking into account the unified projected information structure of the enterprise. "Binding" of various types of local automation systems into a single structure leads to an increase in the start-up period of the complex and an increase in the costs for training the maintenance personnel and for maintenance of spare parts, tools and accessories in the working state.

Content from this work may be used under the terms of the Creative Commons Attribution 3.0 licence. Any further distribution of this work must maintain attribution to the author(s) and the title of the work, journal citation and DOI.
Published under licence by IOP Publishing Ltd

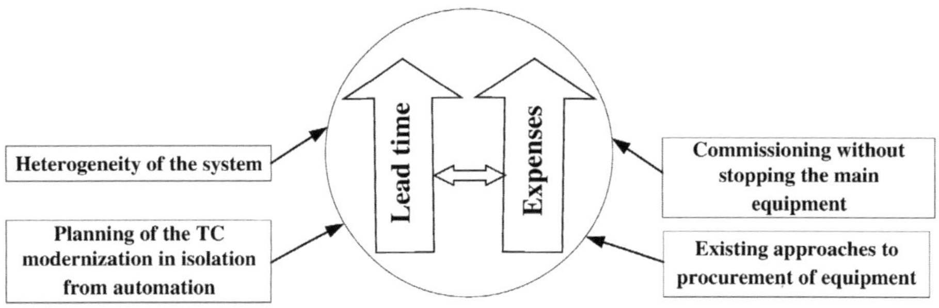

Figure 1. Factors arising during the PP modernization.

Planning TC modernization in isolation from automation. In the current practice, the planning of modernization (reconstruction) of the TC production facility is carried out without taking into account the terms of development of design solutions for technological complex automation. Moreover, the requirements for technological processes and equipment are developed separately and for the means of automation separately as well. This systemic drawback is inherent in the creation of new PP. This is due to the lack of scientific and methodological developments in the joint synthesis of control object and control subsystem.

Of course, the works on design of automation systems should be started as early as possible and, ideally, should be conducted in parallel (together) with the control object [1]. However, in practice, the sequence of designing industrial complexes consists, firstly, in design of technological schemes (objects), and then the control subsystem. From this it follows that the control subsystem is assigned the achievement of the required properties of the entire control system. This can lead either to a significant complication of the control subsystem or even to the impossibility of obtaining the required property of the entire system. In addition, the deadlines for commissioning the entire industrial site are also missed.

In addition, the technical task for the PP modernization (reconstruction, re-equipment) with the introduction of a part of the "new" technological units and sections is done separately from the existing TC (or affects it minimally, taking into account the "link-up" of material flows). The technical task for automated control systems is also done for newly introduced units and sections. However, it is necessary to modernize both control automation systems in terms of newly introduced equipment and the existing automation system, which, as a rule, is not taken into account at the stage of development of the Terms of Reference, and, accordingly, time and financial resources are not allocated.

Also, when choosing new technological equipment, it is necessary to take into account the linkage of electrical equipment, instrumentation, software and the hardware of an existing complex. The choice of software and hardware for new technological areas should be carried out taking into account the already existing similar means or ensuring their integration.

Relevant resources should also be included into development of hardware and software of integration blocks of the existing and new elements of the system.

Start without stopping the main equipment. Often the automation system, newly commissioned within the framework of modernization, is directly connected with the existing technological complex and its control system, for example, an automated dispatch control system. Integration of the newly introduced system with the existing system must be performed on the stopped technological equipment. However, in practice, the approved production plans for the processing of minerals allow the production equipment to be shut down only for the performance of scheduled preventive maintenance (SPM), which is often not enough to perform integration and debugging procedures. Thus, the commissioning of the updated automation systems is done piece by piece, during performance of SPM, competing in time with the work carried out by the maintenance personnel of the production facility. This significantly delays the commissioning period.

Thus, it is more rational to begin modernization of the system from the formation and installation of cabinets with new equipment, taking into account the close integration of the installed equipment with the existing one. Then, step by step (either by units or by technological subsystems), transfer the functions of the old system to the new one with the possibility of parallel operation of both systems. Dismantling of obsolete equipment is done at the end, only after the new system is put into operation.

When the control cabinets are ready and the signal lines are laid from the control panels of the units to the PLC cabinets, the following is performed:

- alternate switching on of temporary control schemes of technological units that can be stopped for commissioning; and it is necessary to provide for the possibility of functioning of a part of the units under the control of the old one, and part of the new system;
- in the time scheme the control of the upgraded units is transferred to the controller, however, for the purposes of monitoring and controlling the units the signal lines of the old system are used; the overall control of the technological complex and other units remains unchanged, the control also remains on the mimic diagram; at this stage the software debugging of the system being put into operation is conducted;
- after software debugging of all units controllers, a transfer to the final control scheme of the units is carried out; all unnecessary signal connections are removed;
- on the readiness of the upper level control (SCADA-system), the dispatching control is transferred from the mnemonic scheme to the workstation;
- after debugging the local control system of each unit, the control of the whole technological complex is transferred with the help of controllers of the updated automated control system.

The use of this kind of phased modernization, on the one hand, requires a lot of time to put the updated system into operation, on the other hand, it allows all routine maintenance of the system modernization to be done without long stops of the process equipment, which positively affects the output of the final product.

Purchase of equipment. In the existing procedures for procuring technological equipment (especially when forming requirements for equipment manufacturers) automation specialists from both the operator and the project organization are often not involved. This leads to the order of different types of local automation systems. When ordering hardware and software automation tools, the cost of the equipment is crucial, which is not always justified. It is necessary to consider other factors: the delivery time, the manufacturer and supplier's warranty, the number of suppliers and so on.

3. Conclusion

Thus, the small involvement of automation specialists, especially at the early stages of development of system-wide design solutions, during the technological complexes modernization of coal-preparation plants and their automation systems may cause a significant rise in the cost of the project and (or) an increase in the timing of its implementation. It is necessary as early as possible, even at the stage of forming the Terms of Reference and the time plan for the modernization of the production facility, to attract automation specialists, which directly relates to the approach of joint synthesis of the control object and the control subsystem [1, 2].

Acknowledgements

The work was carried according to the State Assignment of the Ministry of Education and Science of the Russian Federation No. 8.8611.2017 / 8.9.

References

[1] Ivushkin K A, Myshlyaev L P et al 2014 *Izv. Vuzov. Ferrous Metallurgy* **12** 33–36
[2] Lyakhovets M V, Myshlyaev L P et al 2013 *Proc. of the IX All-Russian Sci. and Practical Conf. in Automation Systems in Education, Science and Production* (Novokuznetsk: SibSIU) pp 52 –54

IOP Conf. Series: Materials Science and Engineering **354** (2018) 012009 doi:10.1088/1757-899X/354/1/012009

Modernization of the automation control system of technological processes at the preparation plant in the conditions of technical re-equipment

M V Lyakhovets[1], K G Wenger[2], L P Myshlyaev[1], M V Shipunov[3], V V Grachev[1], M Yu Melkozerov[4] and Sh A Fairoshin[2]

[1]Siberian State Industrial University, 42 Kirova street, Novokuznetsk, 654007, Russia
[2]"Stroyservis" CJSC, 121 Kuznetsky ave., Kemerovo, 650055, Russia
[3]"Scientific-research Center for Control Systems" LLC, 42 Kirova street, Novokuznetsk, 654007, Russia
[4]"Barzasskoye Tovarischestvo" LLC, 1V Nizhniy Barzas street, Berezovsky, 652421, Russia

E-mail: mail@nicsu.ru

Abstract. The experience of modernization of the automation control system of technological processes at the preparation plant under the conditions of technical re-equipment of the preparation plant "Barzasskoye Tovarischestvo" LLC (Berezovsky) is considered. The automated process control systems (APCS), the modernization goals and the ways to achieve them are indicated, the main subsystems of the integrated APCS are presented, the enlarged functional and technical structure of the upgraded system is given. The procedure for commissioning an upgraded system is described.

1. Introduction

Coal mining is inseparably linked with its processing – enrichment, therefore, together with mines and coal open-pit mines the preparation plants have been long functioning [1-3]. In addition, modern market realities are such, that in the presence of constant demand it is necessary to increase the production capacity of the enterprise. Thus, a modern mining enterprise is a "living" organism, constantly changing its technological and logical structure and, accordingly, automation systems of technological processes. At the preparation plant "Barzasskoye Tovarischestvo" LLC (Berezovsky) the decision has been made to carry out technical re-equipment with increasing the production capacity and qualitative indices of coal enrichment mined at the open-pit mine "Barzassky". At present, the modernization of the technological complex is connected with the modernization of automation systems of the corresponding technological processes.

2. Brief characteristics of the control object

Within the framework of technical re-equipment it is planned to create two sections: a flotation-filtration and drying-furnace ones, and the modernization of the existing technological complex of the concentrate storage. The operating preparation plant is a complex facility with a lot of technological equipment, continuous transport lines with their territorial distribution. Moreover, technological processes are automated, and the task of modernization is, inter alia, in the targeted change of the

Content from this work may be used under the terms of the Creative Commons Attribution 3.0 licence. Any further distribution of this work must maintain attribution to the author(s) and the title of the work, journal citation and DOI.
Published under licence by IOP Publishing Ltd

existing automation system that allows, with minimal material and time costs, the automation of newly introduced equipment and technological processes to be implement without disrupting the overall system.

The automated process control systems (APCS) implemented within the framework of modernization is intended for complex automation of control and management of production and technological processes, equipment of a re-equipped preparation plant [4-7], including the solution of the following additional tasks (taking into account that the existing system fully performs the functions assigned to it and does not require the modernization of any software): centralized control, analysis and display of information on the state of production and technological processes and technological complex equipment of the flotation-filtration and drying furnace sections, a concentrate storage; automation control of technological processes and equipment of the technological complex of flotation-filtration and drying-furnace sections, a concentrate storage.

3. The purpose of system creation

The purpose of modernization is to increase the efficiency of technological complex control of the flotation-filtration section; drying-furnace section; concentrate storage and, as a result, improvement of technical and economic indicators of its functioning: decrease in unit costs for production and increase in the production of a good product; as well as improvement in the quality of commercial products.

The achievement of this goal is provided in the following directions:

1. Automation of information and control functions in solving problems: operative formation and analysis of information on changes in operation modes and the state of technological processes, units and equipment; operative consistent correction of tasks for the operating conditions of technological processes; operative implementation of control decisions and regulation of technological parameters; control, accounting and analysis of violations of technological and production discipline, management effectiveness.

2. Improvement in the reliability of automation system for technological complex control, efficiency and quality of control and management due to: application of modern technical means, methods and algorithms for automatic control, analysis, diagnostics of the state and control of technological processes and equipment; complex and detailed display of information on the state of equipment and units, changes in technological parameters, actions of operational personnel in the system; use as a technical base on the lower level of a system of modern microprocessor controllers with high reliability, long time between failures, easy replacement of failed components, expansion of technical structure, modification of mathware and software; minimization of the number of unreliable electromechanical devices in local systems of logical control and control of individual mechanisms and units.

4. Functional structure of the system

The automated process control system for the flotation-filtration, drying-furnace sections and concentrate storage (APCS FFS) is developed as a component of the integrated automated process control system for the preparation plant (APCS PP) of the preparation plant "Barzasskoye Tovarischestvo" LLC, based on the existing software-hardware means of the operating APCS PP of the technological complex of enrichment of the I and II stages and the loading complex. The APCS FFS is developed taking into account the maximum possible use of the resources of already existing technical means of the automated process control system, unification of software and hardware for the designed APCS FFS and the operating APCS PP of the technological complex of enrichment of the I and II stages.

The integrated functional structure of the integrated APCS PP is shown in figure 1.

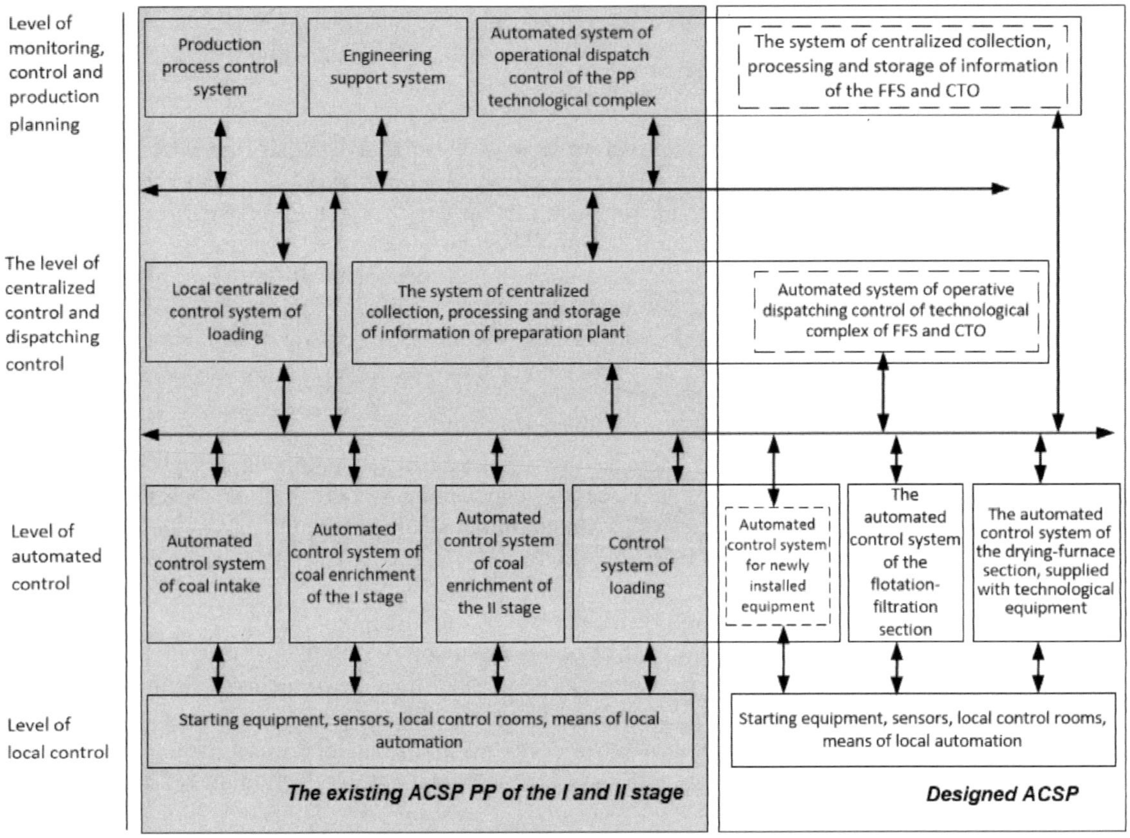

Figure 1. Diagram of the functional structure of the integrated APCS PP of "Barzasskoye Tovarischestvo" LLC.

It consists of the following main systems:

1) the existing APCS PP of the first and second stages, including:

- automated system for the centralized collection, processing and storage of information on PP;
- engineering support system;
- automated control system of production (ACSP);
- automated system of real-time operations control (ASROC) of PP technological complex;
- local centralized control and loading control system;
- automated control systems and technological complexes control (AS TC) for coal intake, enrichment and loading of the I and II stages;

2) the projected APCS FFS, including:

- automated system for the centralized collection, processing and storage of information on the flotation-filtration and drying-furnace sections;
- automated system of real-time operations control of the technological complex of the flotation-filtration and drying-furnace sections (ASROC FFS);
- system of automated control and loading control in the part of newly installed equipment at the concentrate storage;
- automated control systems and technological complexes control of the flotation-filtration section (ACS TC FFS);
- automated control systems and technological complexes control of the drying-furnace section, supplied with the technological equipment, including the local system of centralized control and control of drying units (ACS TC CTO).

IOP Conf. Series: Materials Science and Engineering **354** (2018) 012009 doi:10.1088/1757-899X/354/1/012009

Integration of the existing APCS PP of the first and second stage and the newly created APCS PP FFS and the creation of a complete automated PP technological complex is provided by:

- creation of a single information space in the system of centralized collection, processing and storage of information and in the a automated system of real-time operations control of the plant technological complex, the general discipline of data visualization and dispatching control;
- incorporation into the APCS mathware of control algorithms of technological operation modes and interlocking of equipment at the points of conjugation of the existing technological complex of the I and II stages of the factory and the newly created technological complex of the flotation-filtration, drying-furnace sections and the concentrate storage;
- partial use of the software and hardware of the APCS PP of the first and second stages of the I and II stages for the implementation of the functional systems APCS FFS.

The automated system for centralized collection, processing and storage of information on the technological complex of the flotation-filtration section and ASROC FFS are implemented on the basis of the software and hardware of the existing APCS PP of the first and second stages by expanding its infoware and software.

The designed APCS FFS is a two-level system.

The upper level is a system for the automated system of real-time operations control and technological complexes control of flotation-filtration and drying-furnace sections.

The lower level – automated control systems for technological complexes of flotation-filtration section, drying-furnace section, loading (concentrate storage).

The use of this kind of two-level automation systems has been tested at other coal enterprises and has proven itself well.

5. Technical structure of the system

Scheme of the enlarged technical structure of the ACSP of technological complex for coal loading at the open-pit mine "Barzassky" of "Barzasskoye Tovarishchestvo" LLS at "Barzasskaya" station is shown in figure 2. The implementation of the system is based on microprocessor-based programmable controllers, servers, personal computers and touch panels.

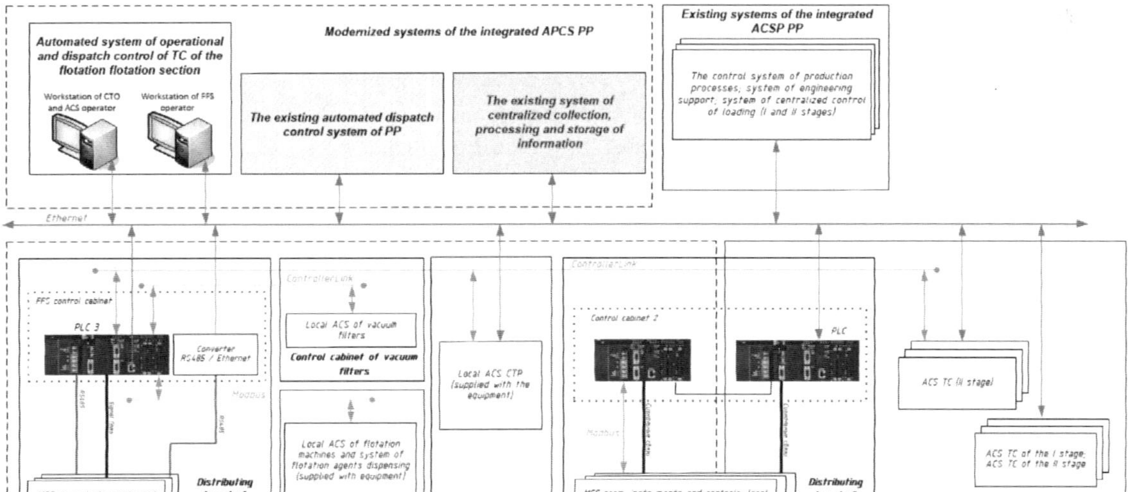

Figure 2. Scheme of the enlarged ACSP technical structure.

Hardware implementation of upper ***level systems***, including ***existing systems of integrated APCS PP*** (control system for production processes; engineering support system; system for centralized control and management of first and second loading; centralized collection, processing and storage of information; automated system of real-time operations control of the technological complex of the factory), as well as the ***designed system APCS FFS*** (automated control system for technological processes of flotation-filtration, drying-furnace sections and concentrate storage) are performed on the basis of servers and personal computers of the following corporations Hewlett-Packard, DELL, HMI-terminals of Schneider Electric Company.

As the basic software of the upper level of the newly introduced APCS FFS, the existing software package Genesis32 from ICONICS v.9.21 is used. The functionality of this application package is sufficient for the performance of all automated functions, the purchase of additional software modules is not required [8].

The subsystem of the ***lower level*** is designed to solve the problems of interconnected control of technological units. In the starting complex it is automated control of the units with observance of the technological regulations of the scheduled start / stop and emergency blocking of the units.

The subsystem is implemented on Omron's CJ2 microprocessor-based programmable logic controllers and provides: reception and processing of discrete signals about the state of equipment and units, generation of discrete signals for units control, realization of the functions of local, remote and automatic equipment control; reception and processing of signals of the measuring information from local systems of measurement and control of technological parameters; it is possible to receive standard analog and discrete signals; the exchange of data between controllers over a specialized Controller Link controller network; data exchange via the Ethernet information network with the upper level workstations of the system; automated or by commands of the dispatcher formation of start / stop commands in accordance with the current state of the process and units, the requirements of the regulations.

Selection of input-output modules, construction of a system for collecting, processing and transmitting information and control actions were performed in accordance with the scheme of the enlarged technical structure of the system (figure 2), the composition of equipment and assemblies of the technological complex of the factory.

The interconnection of the ACSP controllers with the means for measuring and controlling the parameters of technological processes and the state of the equipment is provided by analog (4-20mA) and discrete (24VDC) signals. To transfer control actions to the control circuits of electric drives, "dry contacts" are used. To ensure galvanic isolation of discrete outputs of controllers, Omron G70A-ZOC16-3 quick mounting system is used.

Lower level systems are built using controllers interconnected by means of Controller Link and Ethernet networks, with the transition, if necessary, to a fiber-optic cable for connecting controllers located in other distribution points.

To interface the ACSP controllers with the means for measuring and controlling the parameters of technological processes and the state of the equipment, as well as the executive devices supporting the exchange of information via the RS485 interface (Modbus RTU protocol), the corresponding interface modules are provided in the controllers.

Information exchange between controllers, devices connected via RS485 to the corresponding Omron controller communication modules, with dispatching stations, with servers and with operator HMI terminals is carried out via the Ethernet information network.

6. Commissioning of the system

The ASROC commissioning is performed in three stages.

At the first stage, the modernized system of automated control and technological complex control of the concentrate storage is put into operation.

At the second stage, the automated control system and the technological complex control of the flotation-filtration section is put into operation.

At the third stage, the automatic control system and technological complex control of the drying furnace section are commissioned, which is supplied in a complete set with technological equipment, including a local system for centralized monitoring and control of drying units.

The task of modernization of the existing automation system is complicated by the fact that the PP is not closed for reconstruction and can not stop for a long time, remaining an active production. In this regard, any changes in the structure and functions of the system should be carried out within strictly limited time of planned maintenance and repair stops.

7. Conclusion

The consistent implementation of these design solutions will create an integrated automated technological complex of the preparation plant with close integration of the existing APCS PP of the first and second stages of the newly created APCS FFS.

References

[1] Myshlyaev L P, Kiselev S F, Ivushkin AA et al 2003 *Control Automation of Coal Preparation Plant* (Novokuznetsk: SibGIU) p 304
[2] Sazykin G P et al 2003 *Designing and Construction of a New Generation of Coal Preparation Plants* (Novokuznetsk: SibGIU) p 127
[3] *Industrial Automation Systems* 2006 ed L P Myshlyaev vol 2 (Novokuznetsk: Science) p 483
[4] Grachev V V et al 2015 *Proc. of X All-Russian Sci. Conf. on Automation Systems in Education, Science and Production* (Novokuznetsk: SibGIU) pp 90–95
[5] Grachev V V and Shipunov M V 2014 *Proc. of VI All-Russian Sci. Conf on Automated Electric Drive and Industrial Electronics in the Metallurgical and Mining-fuel Industries* (Novokuznetsk: SibGIU) pp 226–232
[6] Ivushkin K A et al 2015 *Proc. of Int. Sci. Conf. on High Technology of Development and use of Mineral Resources* (Novokuznetsk: SibSIU) pp 203–211
[7] Myshlyaev L P et al 2006 *Proc. of XV Int. Congress on Coal Preparation* (China)
[8] Korovin D E et al 2017 *Proc. XI All-Russian Sci. Conf on Automation Systems in Education, Science and Production* (Novokuznetsk: SibGIU) pp 254–259

IOP Conf. Series: Materials Science and Engineering **354** (2018) 012010 doi:10.1088/1757-899X/354/1/012010

Experience of creating a multifunctional safety system at the coal mining enterprise

V V Reshetnikov[1], K S Davkaev[2], M V Korolkov[2] and M V Lyakhovets[3]

[1]"MMK-Ugol" LLS, 27 1-Teleut street, Belovo, 652607, Russia
[2]"Synergo Soft Systems" LLS, 91a Stroiteley ave., Novokuznetsk, 654005, Russia
[3]Siberian State Industrial University, 42 Kirova street, Novokuznetsk, 654007, Russia

E-mail: lyakhovets_mv@siu.sibsiu.ru

Abstract. The principles of creating multifunctional safety systems (MFSS) based on mathematical models with Markov properties are considered. The applicability of such models for the analysis of the safety of the created systems and their effectiveness is substantiated. The method of this analysis and the results of its testing are discussed. The variant of IFSB implementation in the conditions of the operating coal-mining enterprise is given. The functional scheme, data scheme and operating modes of the MFSS are given. The automated workplace of the industrial safety controller is described.

1. Introduction

The federal rules and regulations in the field of industrial safety "Safety Rules in Coal Mines" approved by Rostekhnadzor Order No. 550 of 19.11.2013 determine the composition of the safety systems included in the MFSS and also define the functions of this system – "monitoring and prevention of the conditions for occurrence of the hazard of geodynamic, aerological and technogenic nature; operative control of conformity of technological processes to the given parameters; application of emergency protection systems for people, equipment and structures" [1]. To solve the problem of providing the function of "operative control of the conformity of technological processes to specified parameters", it is proposed to create an integrated class of automated systems [by analogy with the connection of APCS (automated process control system) – ASODC (automated system for operational dispatch control)]: II MFSS (integrated information multifunctional information safety system) – ASS (automated safety system).

The integrated information multifunctional safety system (II MFSS) is an analog of the ASODC, and this system, as an industrial safety controller, is designed to solve a range of tasks for timely response to emergencies and their prevention. The basis for the allocation of II MFSS from ASODC is:

- priority boost of dangerous situations and accidents over the overall control of technological processes and their separation into a separate category of production tasks;
- information flow in control of complex production exceeded the ability of a person to register minor safety rules failure (SR), which usually together lead to an emergency.

Proceeding from the foregoing, the creation of the MFSS is suitable for complex or dangerous technological industries, for example, a coal mining enterprise.

Content from this work may be used under the terms of the Creative Commons Attribution 3.0 licence. Any further distribution of this work must maintain attribution to the author(s) and the title of the work, journal citation and DOI.
Published under licence by IOP Publishing Ltd

In describing the model of the MFSS and its implementation, it is proposed to use Markovian chains in combination with synergetic models due to the comparative simplicity and clarity of the mathematical apparatus, the high reliability and accuracy of the solutions obtained.

2. Modeling of a multifunctional safety system

Markovian chains are used to synthesize the model of the MFSS. Markovian chains use the general principles of the theory of stochastic processes for analyzing the discrete space of a state of objects group in time. Each object is at any time in one of the states from a bounded set.

The basic postulates of the MFSS model are:

1. Dangerous production – mine – is considered as a synergetic object with many internal states.
2. Transitions from one state to another are regarded as the Markovian process, that is, with independent probabilities.
3. Physical transition of an object from one state to another is displayed in the information space, as a given, expected event.
4. Each event generates a signal.
5. The signal is transmitted to the message by the II MFSS system.
6. The system has negative feedbacks on control, implemented by organizational and technical methods.
7. The task of the MFSS is to prevent the transfer of a dangerous object to the state of emergency.
8. When the object transit into the emergency state, the MFSS should minimize the damage and help restore the object to its normal state.

The model of the II MFSS (figure 1) can be represented in the form of system transitions from one state to another, generating an event and generating a signal; in the initial state of the system – normal operation of technological objects – "normal system operation" (S1); the quantized space of the system state can be represented in the form of such ranked states of severity as: violation of SR (S2), warning (S3), incident (S4), accident (S5). The operator's task is to prevent the system from transition to the "emergency mode of the system" state (S6). This model can be referred to the class of Markovian processes with discrete states and continuous time.

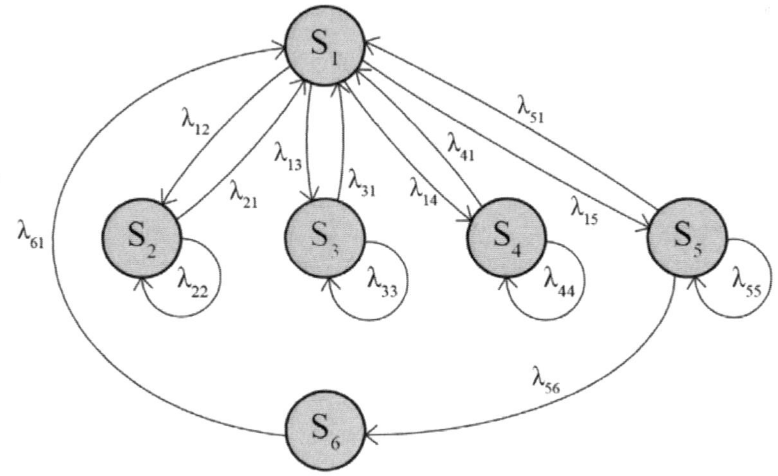

Figure 1. Graphical representation of the II MFSS:
λ_{ij} – probability of transition from the state of S_i to the state S_j.

The main problem in this model is a large number of states of the system and possible transitions between them. To optimize the operation of the system, the provisions of relational algebra are applied. Classification of events and their properties made it possible to identify groups of entities and divide them into sets with small capacities. Optimization of the work of the model is achieved by synchronization of the directories of the II MFSS databases and automated safety systems (ASS).

Classification of events occurs for each ASS according to the following principles:

1. Violation (damage) of the technical process environment (for example – rock fall, methane explosion, fire, flooding, etc.).
2. Violation (damage) of the means of the technical process (equipment), including power supply.
3. Failure of control systems, including damage to communications;
4. Deviation from the specified parameters of the technological process, including unscheduled stops.
5. Errors (including injuries) of personnel, other types of SR violations and warnings.

The event properties are:

1. Priority of the event (Accident, Incident, SR Violation, Confirmation or Prevention)
2. The place and time of the event;
3. Values of the event parameters (for example: the percentage of methane when the maximum permissible concentration is exceeded);
4. Personalization and acknowledgment of the event.

This approach offers the following advantages:

- MFSS directories contain all the events for all safety systems necessary and sufficient for safe operation;
- the system is universal and allows new events to be added;
- the business processes for officials to eliminate the safety threats are specified, as a given reaction to specific events;
- the time required to carry out the necessary instructions for personnel actions in emergency situations is shortened, and the accuracy and concreteness of these instructions increase.

3. Implementation of the multifunctional safety system

In the proposed model of the II MFSS collects and filters data from the ASS for its subsequent processing and storage. ASS can be a separately implemented specialized automated system, or be built into the process control system. The exchange of data from the ASS to the II MFSS is in the form of signals, the signals are converted into event messages. The data collection is organized in the II MFSS both by events (the initiator of the signal sending is the ASS) and by the timer – the external ASS is periodically fetched and signals are generated. The functional scheme of the interaction of systems is shown in figure 2.

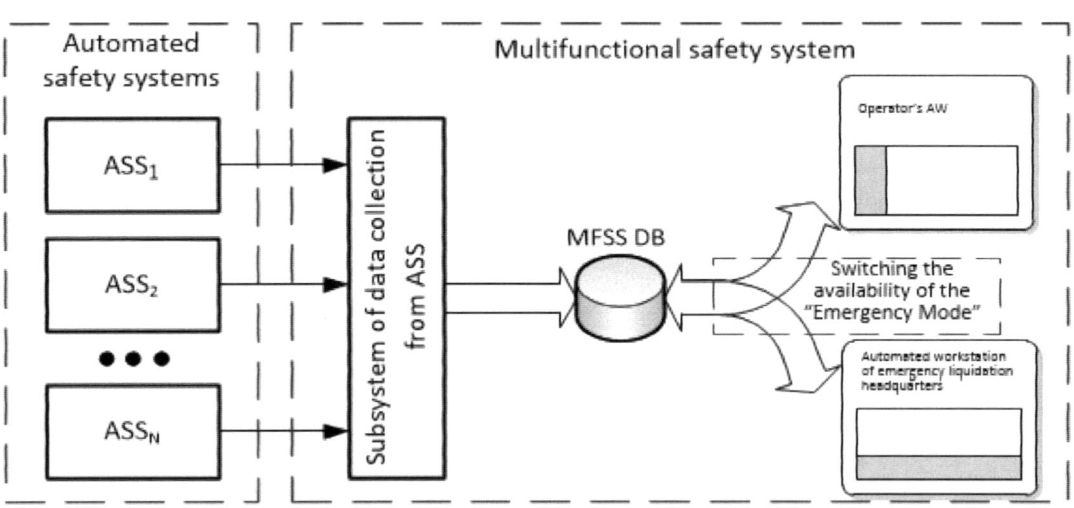

Figure 2. Functional diagram.

Information coming from adjacent automated systems is accumulated in the database (DB) of the MFSS. The data schema is shown in figure 3.

Figure 3. Data Scheme of the MFSS.

Signals are decrypted into messages understandable to the industrial safety controller. The signals are transmitted in the coded form.

The software implementation of automated workplaces is based on the configuration of "1C: Enterprise 8. Industrial safety. Integrated", designed to automate the tasks of ensuring labor safety, industrial, fire and environmental safety at enterprises. The choice of this solution is due to the wide possibilities for automation of the planning, accounting and control processes in various areas of fire and industrial safety, including the ability to keep records of technical devices used at hazardous production facilities, monitor the timeliness of routine maintenance and maintenance for equipment, to register the data of accidents and incidents at hazardous production facilities.

The appearance of the automated workstation (AW) of the operator is shown in figure 4

The system can operate at any time in one of two modes:

Normal operation.

IOP Conf. Series: Materials Science and Engineering **354** (2018) 012010 doi:10.1088/1757-899X/354/1/012010

Under normal operating conditions, the operator receives reports of SR violations, incidents, accidents and from the self-diagnostic subsystem of the MFSS through the event log.

Figure 4. Appearance of the operator's workstation.

The operator must necessarily confirm the message receipt (acknowledge it). If the message is not acknowledged within the specified time period, the system goes into emergency mode. When the message is registered in the system, it gets the status of active (unfinished).

After acknowledging the message, the operator, on the basis of the job description and regulatory documents, must make a decision to eliminate the violation of the SR or incident. When the cause of the event is eliminated, the message is automatically terminated and gets the archive status. It is also possible to terminate the event manually. But in this case, the message will be delayed for the period specified in the system setup, and will appear again, if in fact the factors that caused it are not eliminated.

When the operator enters the system or on demand, he receives a list of pending unfinished messages.

If the time for the completion of the event, that is set during the system setup and depends on the violation correction procedure, expires, this event is completed, and an event with a higher priority is generated instead (for example, a SR violation becomes an incident). In this case, additional automatic notification of the responsible officials from the alert list occurs.

When receiving an oral, telephone, written or other non-electronic message, the operator enters it through a special form in the MFSS.

Emergency operation mode – when the emergency response plan is activated.

In emergency operation mode the operator waits for the emergency mode to end and can generate reports on the occurred events. Emergency mode is used in the following cases:

- when an emergency message is received;
- on the initiative of the responsible head of the accident liquidation or the mining operator;
- if the message is not acknowledged within a specified period of time by the operator or by the person who replaces him.

In the emergency mode, system messages are acknowledged automatically.

4. Technique of the analysis of industrial safety

Accumulated information on messages for a long period makes it possible to calculate the transition probabilities from one state of the system to another, which allows the most probable ways of accident propagation to be predicted.

Analyzing the data it is possible to establish "bottleneck" places of the industrial safety system at the enterprise. That allows the possible accidents to be prevented by organizational and technical methods and safety systems to be optimized by increasing or, conversely, simplifying those that are responsible for the most traumatic and emergency areas of the technological process.

To automate this process it is possible to use the technologies of artificial intelligence systems – expert systems, neutron networks.

5. Conclusion

The economic efficiency of implementing the proposed models makes it possible to optimize the costs for the creation and operation of the safety systems included in the MFSS. Evaluating the effectiveness of technical systems, implementing safety systems, and determining the required "resolving power" of detecting the SR violations in these systems, that is, optimizing the number of signals generated by the system, the redundancy of these systems is eliminated, which leads to a decrease in the cost of acquisition and ownership.

Or the insufficiency of these systems is eliminated, thereby minimizing the costs and time for liquidation of the consequences of non-eliminated accidents. In addition, it is possible to build a digital simulation model of the MFSS enterprise for forecasting and experiments, for optimization, elimination of bottlenecks, and also to train the personnel preventing breakages and the danger from accidents for the real system.

Approbation of the methodology was started at the Chertinskaya-Koksovaya mine in 2016. Currently, the developed system is being implemented at this production facility.

References

[1] Rostekhnadzor Order No. 550 of November 19, 2013 *"On Approval of Federal Norms and Rules in the Field of Industrial Safety "Safety Rules in Coal Mines"*

IOP Conf. Series: Materials Science and Engineering **354** (2018) 012011 doi:10.1088/1757-899X/354/1/012011

About the mechanism of ERP-system pilot test

V V Mitkov and V V Zimin

Siberian State Industrial University, 42 Kirova street, Novokuznetsk, 654007, Russia

E-mail: zimin.1945@mail.ru

Abstract. In the paper the mathematical problem of defining the scope of pilot test is stated, which is a task of quadratic programming. The procedure of the problem solving includes the method of network programming based on the structurally similar network representation of the criterion and constraints and which reduces the original problem to a sequence of simpler evaluation tasks. The evaluation tasks are solved by the method of dichotomous programming.

1. Introduction

The pilot (experimental) projects are designed to test the principle possibility of achieving the goals created by the ERP-system by testing its main functions by a limited number of users. That is, the content of the pilot project is determined by the set of IT services (the functional volume of testing) to be checked, and by the set of users who will be checking the functioning of these services [1].

If the scope of the pilot project is too small, then the fundamentally important functions of the system will not be adequately tested, and the probability of detecting fundamental errors, when fully implemented, will be high. If the subject area is too large, the amount of work on implementation of the pilot project significantly increases, i. e., there will be no necessary speed and flexibility, because of which the pilot design is organized, and its effectiveness will be close to the effectiveness of full implementation.

Let $S = \{S_j \mid j = \overline{1,m}\} = \{\{s_{ji} \mid i = \overline{1,n_j}\} \mid j = \overline{1,n_j}\}$ be the service portfolio of the created ERP-system. Here j – the number of the business process, and i – the number of the service in the business process. The schedule of development of portfolio services, which is the result of the formation and scheduling of service design, is shown in figure 1.

Let us denote the time duration of the service development s_{ji} by $t_{ji} = t(s_{ji})$, and the time of the start of the pilot test by T^p. The time of the end T_{ji} of the service design s_{ji} is determined by the relation

$$T_{ji} = \sum_{k=1}^{i} t_{jk} \quad . \qquad (1)$$

Content from this work may be used under the terms of the Creative Commons Attribution 3.0 licence. Any further distribution of this work must maintain attribution to the author(s) and the title of the work, journal citation and DOI.
Published under licence by IOP Publishing Ltd

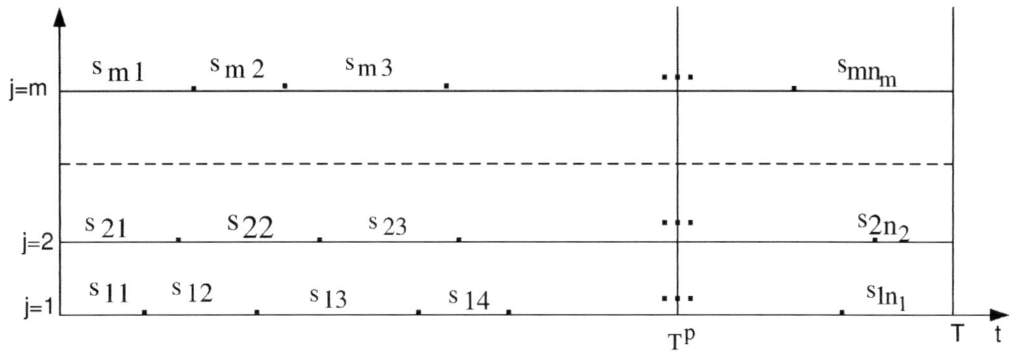

Figure 1. The schedule of the IT services development.

Then the set of services $S(T^p)$, developed by the time T^p of the start of pilot test from which a pilot project can be formed, is determined by the ratio

$$S(T^p) = \{S_j(T^p) \mid j = \overline{1,m}\} = \{\{s_{ji} \mid T_{ji} \le T^p\} \mid j = \overline{1,m}\}\} \subset S \quad (2)$$

To simplify the notation, we enumerate the services of the set $S(T^p)$ so that

$$S(T^p) = \{S_j(T^p) \mid j = \overline{1,m}\} = \{\{s_{ji} \mid i = \overline{1, n_j^p}\} \mid j = \overline{1, n_j^p}\} \quad (3)$$

Let us denote the number of links between the assets of the services s_{ji}, s_{rk} by $q_{ji}^{rk} = q(s_{ji}, s_{rk})$. In accordance with the definition, the pilot test involves the check of correct functioning of the "basic" functions of the system being created by a limited number of "advanced" users. This means that the problem of formation of the scope of the pilot project consists in identification of the services from $S(T^p)$ and the links of these services that must be tested [2-6].

2. Formalization of the problem.
Let us introduce the following variables:

$$x_{ji} = \begin{cases} 1, \text{if the service } s_{ji} \text{ is to be tested,} \\ 0, \text{in the contrary case} \end{cases} \quad (4)$$

and

$$x_{ji}^{rk} = \begin{cases} 1, \text{ if the links of the service } s_{ji} \text{ c } s_{rk} \text{ is to be tested,} \\ 0, \text{ in the contrary case} \end{cases} \quad (5)$$

The total number of links between the services to be checked during the pilot test is described by the function

$$q=\sum_{j=1}^{m}\sum_{i=1}^{n_j^p}\Big(\sum_{r=1}^{m}\sum_{k=1}^{n_r^p} q_{ji}^{rk} x_{ji}^{rk}\Big)x_{ji} \quad . \tag{6}$$

The functional scope of pilot test depends on the amount of costs allocated for its conduct. The costs can include, in particular, the number of standard man-days of competent users, which the company's management team can afford to withdraw from the current core business. Let $\tilde{n}^*$ be the maximum volume of eligible costs. Let us denote the costs required to check the links between the services s_{ji} ñ s_{rk} by $\tilde{n}_{ji}^{rk}=\tilde{n}(s_{ji},s_{rk})$. The total cost of checking the links to be tested will be

$$\tilde{n}=\sum_{j=1}^{m}\sum_{i=1}^{n_j^p}\Big(\sum_{r=1}^{m}\sum_{k=1}^{n_r^p} \tilde{n}_{ji}^{rk} x_{ji}^{rk}\Big)x_{ji} \tag{7}$$

Let us also assume that the management team of each business-process imposes restrictions on the total number of n_j^* business-process services to be tested and on the number of $\bar{n}_j^*$ related business-process services with which the links of the tested services should be checked. Let us consider the following option of formalization of the problem under consideration:

$$q=\sum_{j=1}^{m}\sum_{i=1}^{n_j^p}\Big(\sum_{r=1}^{m}\sum_{k=1}^{n_r^p} q_{ji}^{rk} x_{ji}^{rk}\Big)x_{ji}=\sum_{j=1}^{m}\sum_{i=1}^{n_j^p} q_{ji}x_{ji}=\sum_{j=1}^{m} q_j \to \max \quad , \tag{8}$$

$$\tilde{n}=\sum_{j=1}^{m}\sum_{i=1}^{n_j^p}\Big(\sum_{r=1}^{m}\sum_{k=1}^{n_r^p} \tilde{n}_{ji}^{rk} x_{ji}^{rk}\Big)x_{ji}=\sum_{j=1}^{m}\sum_{i=1}^{n_j^p} c_{ji}x_{ji}=\sum_{j=1}^{m} c_j \le c^* \quad , \tag{9}$$

$$\sum_{i=1}^{n_j^p} x_{ji} \ge n_j^*,\ j=\overline{1,m} \quad , \tag{10}$$

$$\sum_{r\neq j}\sum_{r=1}^{n_r^p} x_{ri} \ge \bar{n}_j^*,\ j=\overline{1,m} \quad . \tag{11}$$

The task is to find such $x_{ji},\ j=\overline{1,m},\ i=\overline{1,n_j^p}$, and $x_{ji}^{rk},\ j,r=\overline{1,m},\ i,k=\overline{1,n_j^p}$ (that is, a subset of IT-services and such subsets of their links) that would maximize the number q of the checked links between the services being tested at the given allowed costs $\tilde{n}^*$ and constraints (10) and (11).

3. The scheme of problem solving

To solve the problem (8) – (11), we use the method of network programming, which assumes a sequential solution of the chain of evaluation tasks, formed on the basis of a structurally similar network representation of the objective function and constraints [7-8].

Then for the problem (8) – (11) the chains of evaluation tasks can be realized in three stages:

1. A sequential solution for each service of each task process of determining the links to be checked:

$$q_{ji} = \sum_{r=1}^{m} \sum_{k=1}^{n_r^p} q_{ji}^{rk} x_{ji}^{rk} \to \max \tag{12}$$

$$c_{ji} = \sum_{r=1}^{m} \sum_{k=1}^{n_r^p} c_{ji}^{rk} x_{ji}^{rk} \le c^*, j = \overline{1,m}, i = \overline{1,n_j^p}, \tag{13}$$

$$\sum_{i=1}^{n_j^p} x_{ji} \ge n_j^*, j = \overline{1,m}, \tag{14}$$

$$\sum_{r \ne j} \sum_{r=1}^{n_r^p} x_{ri} \ge \bar{n}_j^*, j = \overline{1,m}. \tag{15}$$

A structurally similar network representation of functions q_{ji} and c_{ji} is given in figure 2.

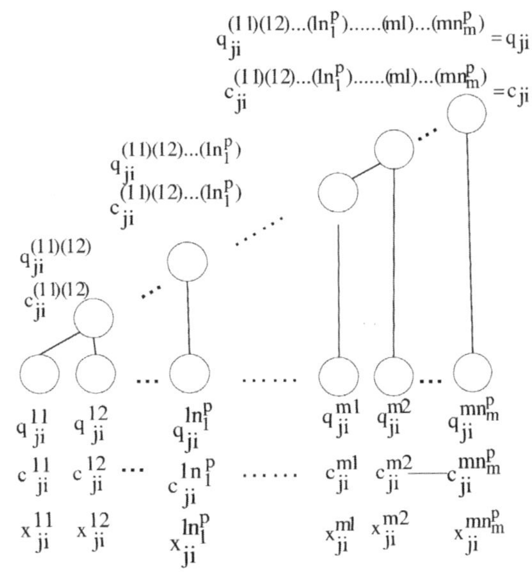

Figure 2. Network representation of functions q_{ji} and c_{ji}.

1. Sequential solution m ($j = \overline{1,m}$) to the tasks:

$$q_j = \sum_{i=1}^{n_j^p} q_{ji} x_{ji} \to \max \tag{16}$$

$$\tilde{n}_j = \sum_{i=1}^{n_j^p} c_{ji} x_{ji} \le \tilde{n}^*, \; j = \overline{1, m} \quad , \tag{17}$$

$$\sum_{i=1}^{n_j^p} x_{ji} \ge n_j^*, \; j = \overline{1, m} \tag{18}$$

$$\sum_{r \ne j} \sum_{r=1}^{n_r^p} x_{ri} \ge \bar{n}_j^*, \; j = \overline{1, m} \quad . \tag{19}$$

A structurally similar network representation of functions q_j and c_j is given in figure 3.

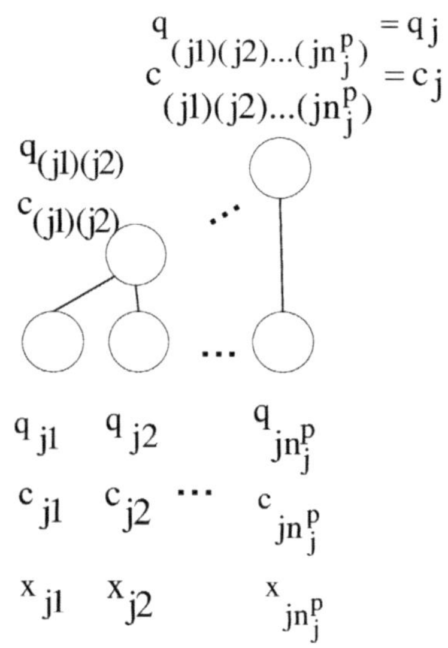

Figure 3. Network representation of functions q_j and c_j.

2. A sequential solution (m-1) to the evaluation task for determining the solution to the task:

$$q = \sum_{j=1}^{m} \sum_{i=1}^{n_j^p} q_{ji} x_{ji} \to \max \tag{20}$$

$$c = \sum_{j=1}^{m} \sum_{i=1}^{n_j^p} c_{ji} x_{ji} \le c^* \quad . \tag{21}$$

A structurally similar network representation of functions q and c is given in figure 4.

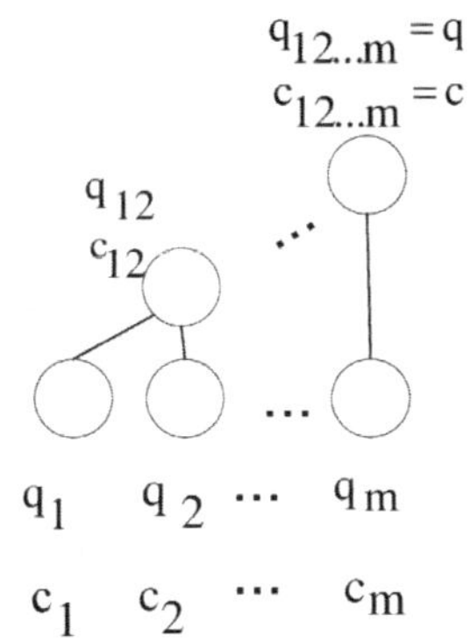

Figure 4. Network representation of functions q and c.

4. Conclusion

The solution obtained in accordance with the above presented scheme is approximate. To find the global optimum, the branch-and-bound method can be used, in which the values q of the found approximate solution can be used as bounds [9-10].

References

[1] Zimin V V et al. 2013 *Fundamentals of Services Lifecycle Control in Computer Science and Automation Systems (best practices ITIL)* (Kemerovo: Kuzbassvuzizdat) p 500
[2] Burkov V N and Novikov D A 1997 *How to Manage Projects: Scientific and Practical Publication* (Moscow: SINTEG – GEO) p 188
[3] Phillips J J, Bothell T W and Snead G L 2003 *The Project Management Scorecards* (Amsterdam: Elseiver)
[4] Zimin V, Burkova I and Zimin A 2017 *Models and Mechanisms of Lifecycle Control of IT Services* (LAP LAMBERT Academic Publisging RU) p 337
[5] Wysocky R K, Beck R and Crane D B 2000 *Effective Project Management* (New York: John Wiley & Sons)
[6] *Mechanisms of Management* 2011 ed by D A Novikov (Moscow: LENAND) p 192
[7] Burkova I V 2012 *The Method of Network Programming in Project Management Tasks* cand. thesis (Moscow) p 40
[8] Burkova I V 2009 *Automation and Telemechanics* **10** 15–21
[9] Wagner G 1972 *Fundamentals of Operations Research* (Moscow: Mir) p 198
[10] Korbut A A, Siga I Kh and Finkelstein Yu Yu 1977 *Math. Operation Forsch. Staist. Ser. Optimization* vol **8 2** 253–280

IOP Conf. Series: Materials Science and Engineering **354** (2018) 012012 doi:10.1088/1757-899X/354/1/012012

Analysis of the environmental and economic indicators of the industrial enterprise

V G Mikhailov[1] and T V Kiseleva[2]

[1]T. F. Gorbachev Kuzbass State Technical University, 28 Vesennyaya street, Kemerovo, 650000, Russian
[2]Siberian State Industrial University, 42 Kirova street, Novokuznetsk, 654007, Russia

E-mail: mvg.eohp@kuzstu.ru

Abstract. In the paper the features of the analysis of the environmental and economic indicators of the industrial enterprise are considered. The purpose of the study is to improve the system of environmental and economic analysis at the enterprise for more accurate forecasting of its main environmental and economic indicators. The study of the main approaches to the implementation of environmental and economic analysis based on the corresponding systems of indicators with identification of the most significant factors was carried out. The main result of the study is the choice of a system for analyzing the environmental and economic indicators, maximally oriented to a specific enterprise, taking into account its production specific features. The practical significance of the study consists in the selection of an adequate system of indicators at enterprises to improve the effectiveness from preparation of an environmentally safe management decision.

1. Introduction

One of the modern trends in the effective operation of an enterprise is the consideration of the environmental factor in the implementation of management decisions [1 - 3] in the conditions of a multi-level management system – from the workplace of the production unit to the Director General. Figure 1 presents a simplified scheme for the development of an effective environmental-friendly management decision.

The first block, presented in figure 1, includes the choice of the system of environmental and economic indicators, which depends on many factors, including the size of the enterprise and its negative environmental impact, industry, availability of reliable data, etc.

In the second block, the analysis of the environmental and economic indicators is performed, in which its purpose is determined, the tasks and the procedure are formed.

The third block is directly linked to the forecasting of environmental and economic indicators, the results of which are used to justify and support the adoption of an effective environment friendly management decision.

The result of the fourth block includes the entire range of control actions aimed at the internal and external environment of the enterprise, including:

- formation of a production program taking into account environmental constraints;
- improvement of the organizational structure of the environmental service;
- implementation of technological solutions aimed at minimizing the negative impact on the environment;

Content from this work may be used under the terms of the Creative Commons Attribution 3.0 licence. Any further distribution of this work must maintain attribution to the author(s) and the title of the work, journal citation and DOI.
Published under licence by IOP Publishing Ltd

- organization of procurement of raw materials and materials based on environmental safety requirements, etc.

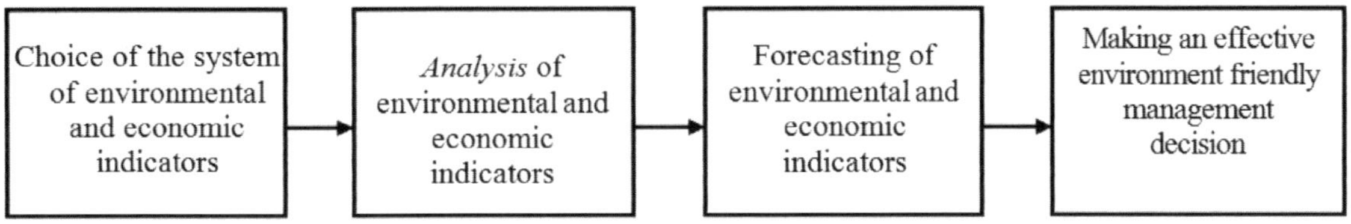

Figure 1. Scheme for the development of an effective environment friendly management decision.

In this regard, for the adequate analysis and forecasting, the choice of a system of environmental and economic indicators is of great importance.

2. Material and methods

The object of the study is the environmental and economic analysis of large industrial enterprises, differentiated according to the areas of environmental policy implementation.

The subject of the study is a system of environmental and economic indicators that is maximally adapted to the specific functioning enterprise.

The study is based on the analysis of domestic and foreign literature devoted to the problems of environmental and economic analysis of enterprises and other territorially distributed environmental and economic systems. Particular attention is paid to approbation of the chosen system of environmental and economic indicators on these large industrial enterprises.

To approbate the proposed methodology, statistical processing of environmental data of an industrial enterprise was carried out on the basis of official environmental reporting forms. The elements of system analysis and results obtained by specialists in the field of environmental and economic analysis of objects of different levels were also used in the work.

3. Results and discussion

There are various approaches to solving this problem, for example, a number of researchers note that in calculating environmental and economic indicators, the methodology used to calculate the economic damage from environmental pollution is of great importance [4, 5]. In the article [6] the authors considered the existing approaches to the assessment of environmental and economic damage, including in the context of economic losses in the sphere of forest use.

In work [7] the technique for calculation of the efficiency of complex use of material resources, based on the system of indicators reflecting social-economic and environmental efficiency, is offered. The group of indicators of economic efficiency includes the following elements:

- volume of sold products produced using wastes [8];
- the amount of profit from the sale of waste and products produced with their use;
- economic efficiency of waste utilization.

Direct analysis of environmental and economic indicators is also a key stage in the making of an effective management decision [9], for example, a number of authors note that the environmental and economic analysis of production is an important part of the environmental management system at macro and micro levels [10]. In work [11] the improved technique of the environmental and economic analysis of environment protection activity of the enterprise for a substantiation of increase of its efficiency is considered. At the same time, the author supplemented the traditional system of principles of economic analysis of the enterprise with specific indicators:

- capital productivity of the main production assets of environment protection purposes, taking into account the amount of prevented environmental damage per one ruble of the cost of environmental protection equipment;

- resource intensity of products – through the specific weight of their consumption in value terms in the cost of production.

In study [12], the calculated indicators were developed for carrying out environmental and economic analysis at the facilities of housing and communal services and incorporated into the following blocks:

- indicators of the state and efficiency of the use of fixed assets, including those involved in environmental protection measures;
- environmental and economic indicators that characterize the impact on the production process, the performance of work, the provision of services;
- environmental and economic indicators that reflect the impact of environmental and economic processes on the financial performance of economic entities;
- coefficients of return on equity, adjusted for the environmental factor;
- turnover indicators.

Based on the study of approaches to the problem of effective environmental and economic analysis, a system of environmental and economic indicators was developed [13], which was tested at the enterprises of "Azot" KJSC and "Koks" PJSC, which have a significant negative environmental impact (NEI). The most important are the statistical dependencies of the level of risk [14], charges for the NEI and the compensation coefficient of economic losses from the current costs of environmental protection. The information base for determining these indicators is the statistical reporting form No. 4-OS "Information on current costs for environmental protection and environmental payments".

The level of risk or the specific weight of the fee for excess NIE in the total amount of payment is determined by formula (1):

$$\mathrm{RL} = \sum_{k=1}^{K}\sum_{l=1}^{L}\frac{\mathrm{Ple}_{kl}}{\mathrm{P}_{kl}}100\ \%, \tag{1}$$

where k is the type of pollutant or hazard class of production and consumption waste; l is an element of the environment; K – total number of types of pollutants or hazard classes of production and consumption wastes; L is the total number of elements of the environment; P_{kl}, Ple_{kl} – respectively, the total amount of payment and payment for the limit-exceeding NIE of the k-th type of pollutant or hazard class of production and consumption waste per l-th element of the environment, million rubles .

Figure 2 shows the correlation field between the current costs of environmental protection and the risk level at the enterprise of "Azot" KJSC. Each point on the analyzed correlation fields displays a calendar year.

Figure 2 shows that the dependence is non-linear, in particular, until 2013 (the points on the left side of the diagram) there is an inverse relationship, when the increase in current costs from 23.8 to 2187.5 million rubles leads to a decrease in RL from 39.59 to 0%. From 2013 (the right side of the diagram) the further increase in current costs is accompanied by a significant increase in RL to 92.58%, which is explained by a sharp increase in payments for limit-exceeding pollution of water sources and may be caused by large-scale technical re-equipment of treatment facilities, which is planned to be completed in 2017.

The forecast of the risk level, carried out with the help of a simple extrapolation, provided that the enterprise environmental policy did not changed during the forecast period, showed that further increase in current costs will not lead to the reduction in the risk level, the value of which will fluctuate about 100%. This situation characterizes insufficiently effective use of current costs for environmental protection and requires a search for a mechanism of their distribution to enhance the company environmental and economic security.

Figure 3 shows the correlation field between the current costs of environmental protection and the payment for NEI by the enterprise "Azot" KJSC.

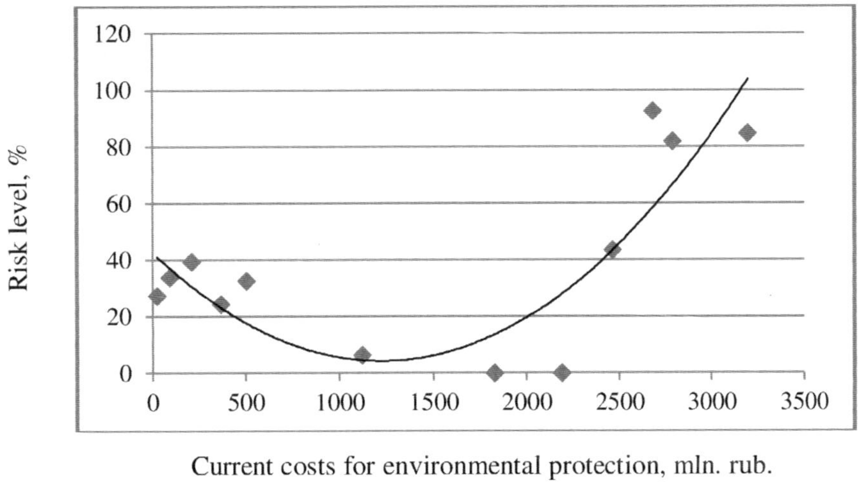

Figure 2. Correlation field and empirical line of regression of the dependence of the risk level on the current costs of environmental protection at the enterprise "Azot" KJSC.

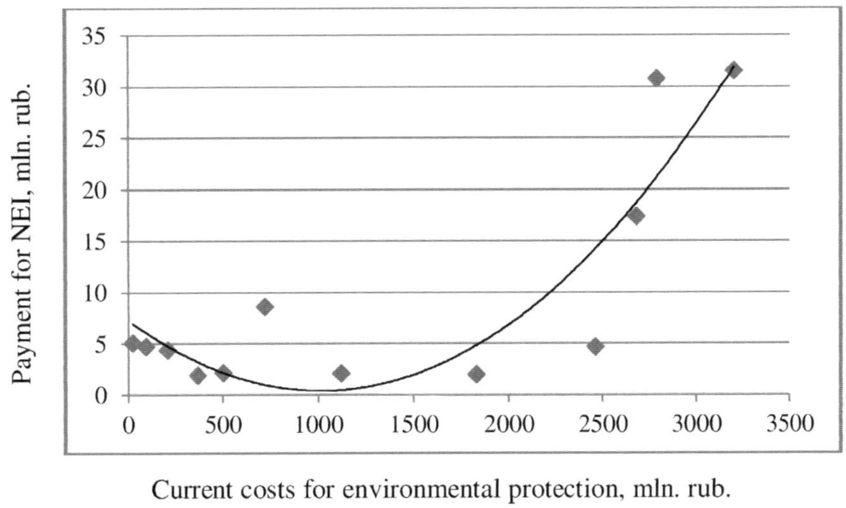

Figure 3. Correlation field and empirical line of regression of the dependence of payment for NEI on the current costs for environmental protection at "Azot" KJSC.

Analysis of figure 3 also shows different conditions for implementing the environmental policy of the enterprise. In the lower part of the correlation field, it can be seen that a significant increase in current costs (from 24 million rubles to about 2.5 billion rubles) led to a slight fluctuation in the payment (from 5 to 2 million rubles). The upper part of the correlation field shows that the further increase in current costs is accompanied by a sharp rise in fees (up to 31.51 million rubles) due to excess pollution of water sources. *The forecast of the NEI payment* is consistent with the conclusions drawn from the analysis of figure 2. In particular, the current costs intended to reduce the negative impact under these conditions do not fulfill this function and with the forecast of their increase to 3.5 billion rubles. The payment for negative impact will also increase and will amount to more than 40 million rubles.

To assess the environmental and economic efficiency of environmental protection, an indicator such as the economic loss compensation coefficient (CCel), which for a particular enterprise is calculated using formula (2), can be used:

$$\text{CCel} = \sum_{k=1}^{K}\sum_{l=1}^{L}\frac{P_{kl}}{\text{EL}_{kl}}100\ \%, \tag{2}$$

where EL_{kl} – economic damage from negative impact of the *k*-th type of pollutant or hazard class of production and consumption wastes on the *l*-th element of the environment, million rubles.

Figure 4 shows the correlation field of the dependence of the economic loss compensation factor on the current environmental costs of the "Koks" PJSC.

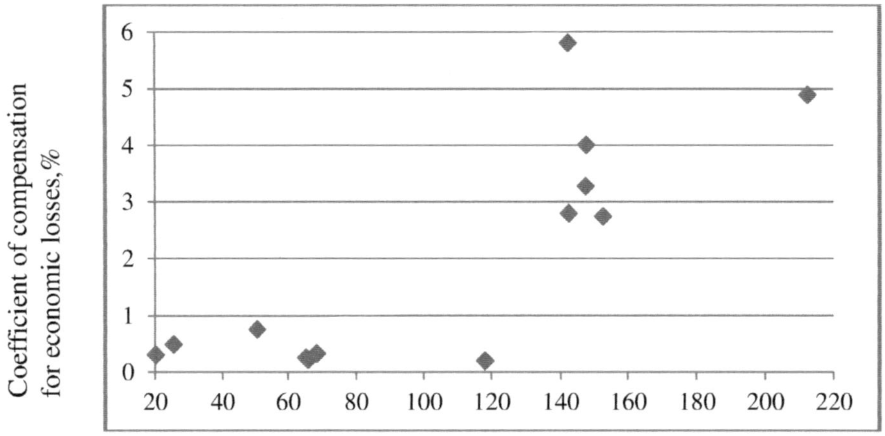

Figure 4. Correlation field of the dependence of the economic loss compensation coefficient on the current environmental costs of "Koks" PJSC.

From figure 4 it can be assumed that environmental activities at the enterprise were implemented under different conditions. The lower part of the correlation field characterizes a significant increase in current costs (from 20 to 69 million rubles), at which CCel is below 1% and almost unchanged. The upper part of the field also shows a sharp increase in current costs (from 142.6 to 213.2 million rubles), but there is a significant increase in CCel (from 2.74 to 5.8%). This situation is explained, in particular, by the commissioning of a closed water cycle, which led to a decrease in the magnitude of economic damage. The general trend of CCel is positive, and with an increase in current costs for environmental protection to 250 million rubles its *predicted value*, also obtained by simple extrapolation, should be more than 4%.

The analysis and forecasting [15, 16] of the considered environmental and economic indicators are necessary for the formation of effective management decisions aimed at improving the quality of current costs. The improvement of the system of their distribution for specific areas of use is of great importance here, for example, using the method of expert assessments. The projected increase in the level of risk and NEI payment also motivates enterprises to improve the efficiency of the organizational and economic mechanism for environmental management, in particular, for "Azot" KJSC it is advisable to plan a production program taking into account environmental constraints.

4. Conclusions

As a result of the study the following conclusions can be drawn:

- a simplified scheme for the development of an effective environment friendly management decision with interpretation of individual blocks was developed;
- a system of environmental and economic indicators was selected based on the analysis of literature sources;
- proposed environmental and economic indicators based on official data and adapted to the specifics of enterprises;
- correlation fields and empirical regression lines for the dependence of the main environmental and economic indicators on current costs for environmental protection are constructed;
- an attempt is made to predict the main environmental and economic indicators taking into account the "non-linearity" of the environmental policy implementation by the enterprise;
- proposals have been worked out for the system of support for making an effective environment friendly management decision.

References

[1] Burkov V N and Burkova I V 2014 *Automation and Remote Control* vol 75 **3** 470–480
[2] Burkov V N and Burkova I V 2015 *Game Theory and Applications* **17** 17-36
[3] Zolotukhin V M, Gogolin V A, Yazevich M Yu, Baumgarten M I and Dyagileva A V 2017 *IOP Conference Series: Earth and Environmental Science* **50** 012027
[4] Mikhailov V G, Golofastova N N, Galanina T V, Koroleva T G and Mikhailova Ya S 2017 *IOP Conference Series: Earth and Environmental Science* **50** 012038
[5] Galanina T V, Baumgarten M I, Mikhailov V G, Koroleva T G and Mikhailov G S 2017 *IOP Conference Series: Earth and Environmental Science* **50** 012030
[6] Zhivotiagina N I et al 2013 *Scientific Journal of KubSAU* **88 (04)**
[7] Dorogovtseva A A, Erygina A V and Dorogovtsev A P 2012 *Economic and Social Changes: Facts, Trends, Forecast* **2(20)** 143–149
[8] Efimov V I, Sidorov R V and Korchagina T V 2015 *Coal* **12** 73–76
[9] Bubnova M B and Ozaryan Y A 2016 *Journal of Mining Science* vol 52 **2** 401–409
[10] Redina M and Khaustov A 2014 *Proc. of SPE Russian Oil and Gas Exploration and Production Technical Conference and Exhibition* (Moscow) vol 1 pp 355–371
[11] Nikitin VN 2006 *Perfection of the Methodology of Environmental and Economic Analysis of the Environmental Protection Activity of the Enterprise*: cand. thesis (Ekaterinburg) p 159
[12] Zakalyukina E V 2012 *Bulletin of the Chelyabinsk State University. – Economics issue* 8 (262) **6** 87–90
[13] Mikhailov V G, Kiseleva T V, Karasev V A, Mikhailov G S and Skukin V A 2017 *IOP Conference Series: Earth and Environmental Science* **66** 012008
[14] Mikhailov V G , Koryakov A G and Mikhailov G S 2015 *Journal of Mining Science* vol 51 **5** 930-936
[15] Trofimov V B and Kulakov S M 2011 *Optoelectronics, Instrumentation and Data Processing* vol 47 **3** 313–320
[16] Mikhailov V G, Kiseleva T V, Bugrova S M, Muromtseva A K and Mikhailova Ya S 2017 *E3S Web of Conferences* **21** 02004

Management mechanisms for development of personnel professional competencies at a loading and haulage company

S V Shishkina[1], Yu D Pristupa[1], L D Pavlova[2] and V N Fryanov[2]

[1]"United Production and Transport Department of Kuzbass" LLC, 13 Kirova ave., Leninsk-Kuznetsk, 652518, Russia
[2]Siberian State Industrial University, 42 Kirova street, Novokuznetsk, 654007, Russia

E-mail: ld_pavlova@mail.ru

Abstract. The necessity for development a management mechanism for development of personnel professional competencies at a loading and haulage company in order to ensuring the compliance of competencies with labor functions, regulated by the current professional standards, is substantiated. A functional diagram of the organizational and technical system of the loading and haulage company was developed, that includes the interaction mechanism between the personnel and the objects of the main production. The problem of algorithm development to assess the correspondence of communicative competencies and labor functions of loading and haulage companies, ensuring the achievement of the management goal with a minimum risk of an emergency situation or an accident, was set and solved. The idea of management problem solving consists in the operational development and implementation of anticipatory measures at each deviation in personnel actions from normative or target indicators. The dependence of the risk of imbalance of communicative competences and labor functions in a given period of time is established. The effectiveness of the developed and implemented mechanism is confirmed by the positive dynamics of the test results.

1. Introduction

The necessity for development of a management mechanism for development of the personnel professional competencies at a loading and haulage company company (LHC) follows from the need to ensure that the competencies of employees are in accordance with the labor functions regulated by the existing professional standards [1, 2, etc.].

Based on the results of the analysis of the target function, structure and activities of the LHC, the company can be identified as a complex organizational and technical system (OTS), the main functions of which are to provide loading and unloading and transport-expediting services for mines, open-pit mines, processing plants, current repair of railway tracks and adjustment of own diesel locomotives.

In accordance with the specifics of the company in the OTS structure three interacting subsystems can be distinguished: the main production, industrial and social infrastructures. The main production of the company includes the production processes of the company, during which materials or raw materials are transformed into products. In the context of LHC these processes are the storage, processing, loading and transportation of coal in wagons for sale to consumers.

The production infrastructure includes a complex of workshops, facilities and services that provide the necessary conditions for the main activity of the company: the rolling stock service, the track service, the material and technical supply department, the communication section, the production

Content from this work may be used under the terms of the Creative Commons Attribution 3.0 licence. Any further distribution of this work must maintain attribution to the author(s) and the title of the work, journal citation and DOI.
Published under licence by IOP Publishing Ltd

control and labor safety service, the production and technical department, department of the automated process control system (APCS).

The social infrastructure unites the divisions of the company, which provide for the sanitary and hygienic working conditions and social needs of workers, the level of their professional training, material support, social guarantees, personnel policy.

The detailed information on the conditions and features of the LHC functioning on the example of "United Production and Transport Department of Kuzbass" LLC is described in the works of the authors [3-6].

The efficiency and safety of functioning of all subsystems is ensured by the level of professional competence of personnel. The personnel policy of the LHC consists in the selection of personnel, assessment of its professional competencies, placement and training, preventive work, continuous monitoring of personnel actions in order to improve and develop occupational safety and production systems.

However, based on the results of the analysis of the dynamics and causes of defects and hazardous production situations, "United Production and Transport Department of Kuzbass" LLC for the period 2000-2014 detected 168 incidents, including:

- hitting people on the railways (24%);
- bursting open the switch (21%);
- derailing of rolling stock (21%);
- collisions of rolling stock (14%);
- other (20%).

It is established that one of the main causes of these incidents are violations of technological regimes, labor and production discipline by workers, that is, an insufficient level of professional competence is confirmed [7].

In this regard, the actual task is to create a mechanism for managing the development of personnel professional competencies at a loading and haulage company.

2. Setting and solving the problem of managing the development of personnel professional competencies

To set and describe the problem of managing the development of personnel professional competencies while minimizing the risks of emergency situations and incidents, the conditions for interaction of employees with the elements of production infrastructure in accordance with the target functions of "United Production and Transport Department of Kuzbass" LLC are considered.

According to professional standards and technological maps, labor functions and actions are systematized for the main activities of the company.

During performance of works, processes and operations, due to the interaction of workers and elements of the production infrastructure, there is an accumulation of inconsistencies in the actions of personnel with the requirements for the operation of infrastructure elements because of the influence of the following factors and events:

- wear out of equipment;
- deviation of the level of engineering solutions from industrial safety requirements and technical instructions for the operation of machinery and equipment;
- unsatisfactory operational control of compliance of actual technological parameters with regulatory ones;
- noncompliance of actions of workers to regulatory technological regimes (violation of technological discipline);
- imperfection of the system of preventive measures to prevent the development of production violations and hazards;
- skills mismatch of an employee when using production resources.

Accumulation of noncompliance of personnel actions with the requirements of infrastructure elements operation for a given time (shift, day, month, year) contributes to the increase in the risk of incidents, pre-emergency situations, occupational diseases, trauma of workers, which is accompanied by negative economic and social consequences for the company in whole. To solve this problem it is possible to use the model of conformity of communicative competences and labor functions of loading and haulage companies.

Three interacting subsystems of the OTS and a set of proactive preventive measures are **given** (figure 1). Personnel and infrastructure elements in the process of performing labor functions interact when performing control actions taking into account the state and changes in the external environment. At the same time, as part of the liquidation plan for accidents and technological maps for the implementation of certain processes and operations, implementation of pre-existing measures previously developed and approved in accordance with the established procedure is carried out.

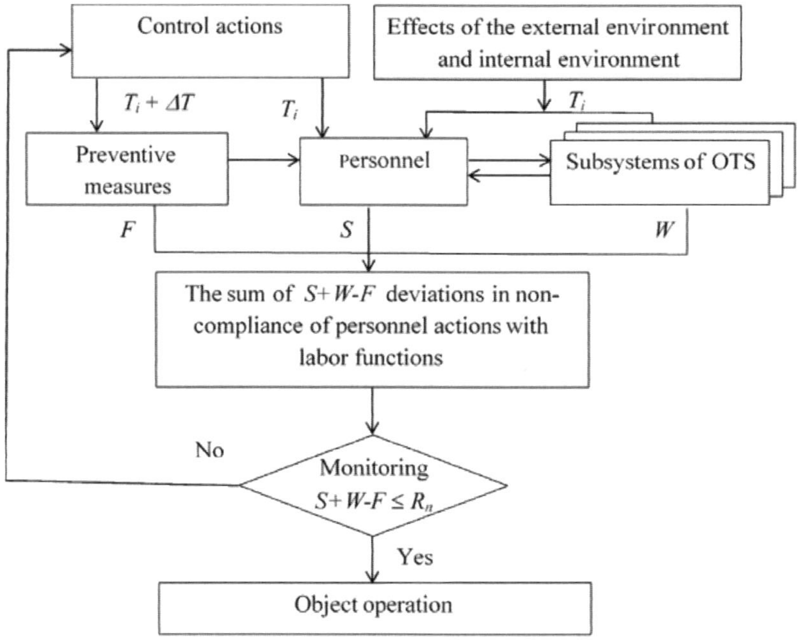

Figure 1. Scheme for processes management of subsystems interaction in the structure of OTS of a loading and haulage company.

It is **required** to establish the dependence of the risk of imbalance of communicative competencies and labor functions during interaction for a given time period $\Delta T = T_n - T_i$ of personnel *S*, infrastructure elements *W* and preventive measures *F*, that is,

$$S + W - F \leq R_n, \quad (1)$$

where T_i – the start time of the process or operation; T_n – the end time of the process or operation; R_n – acceptable risk, is established in accordance with the objective management function of OTS and "Methodological Recommendations ..." [8].

Initial conditions and limitations. The state of the control object in the initial period T_i is assumed to be stable. This is fixed by the results of production control during the period $T< T_i$ and the current monitoring (figure 1). During the period $\Delta T = T_n - T_i$ the professional level and competences of the personnel can be changed by means of raising the level of skills, instructing, changing the working conditions of the personnel. The state of infrastructure facilities also does not remain constant due to wear out of equipment, technical re-equipment of infrastructure elements. Control actions in the form of preventive measures can be developed and implemented step-by-step or discretely according to the recommendations of controlling agents-centers.

Solution to the problem. The idea of the problem solving consists in the continuous improvement of the social safety management system through the operational development and implementation of preventive measures at each deviation of personnel actions and actual operating modes of infrastructure elements from normative or target indicators.

The solution of the problem in general form is as follows:

$$\sum_j\{\sum_{T_i}^{T_n}(p_1(s_1-f_{s_1})+p_2(s_2-f_{s_2})+p_3(s_3-f_{s_3})+\dots+p_m(s_m-f_{s_m}))$$
$$+\sum_{T_i}^{T_n}(q_1(w_1-f_{w_1})+q_2(w_2-f_{w_2})+q_3(w_3-f_{w_3})+\dots+q_v(w_v-f_{w_v}))\}\le R_n\,,$$

where p_m – the risk of an emergency when the personnel actions deviate from the requirements of labor functions when the m-th operation of the j-th service is performed; s_m – the deviation of the personnel competence from the requirements of labor functions when performing the m-th operation of the j-th service; f_{s_m} – measures to prevent the deviations of personnel actions from the requirements of labor functions when the performing the m-th operation of the j-th service; q_v – the the influence of the v-th deviation of actual operating modes of the infrastructure elements from the normative ones on the risk of an emergency situation or an incident of the j-th service; w_v – the v-th deviation of the actual operating modes of the infrastructure elements from the normative ones of the j-th service; f_{w_v} – measures to prevent the v-th deviation of the actual operating modes of the infrastructure elements from the normative ones of the j-th service.

A qualitative schedule for changing the accumulated deviations of inconsistencies in the actions of personnel to the requirements of labor functions in the j-th service is shown in figure 2.

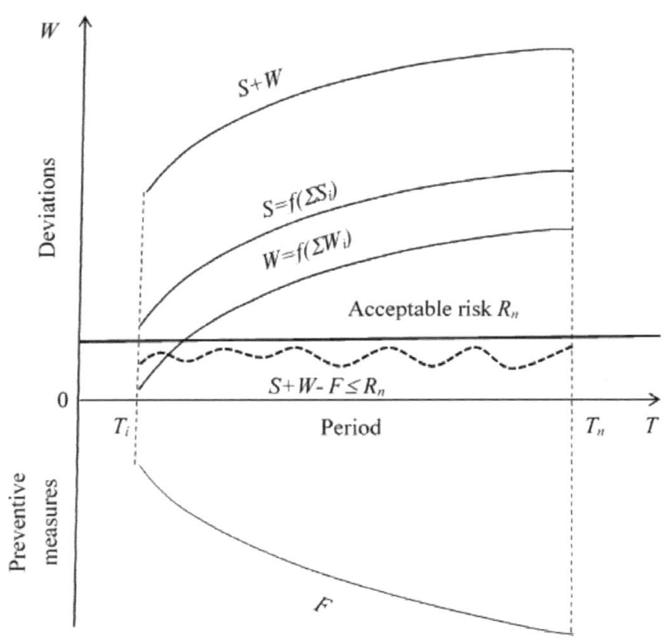

Figure 2. Diagram of accumulation of deviations in the personnel actions from the labor functions regulated by professional standards.

The upper graphs in figure 3 show the changes in the total deviation of non-compliance of personnel actions S with the requirements of labor functions and the deviations of the non-compliance of the actual results of the labor functions performance W from those regulated by professional standards and technological maps. At simultaneous performance of multiple labor functions with deviations from the processes and operations indicated in the technological maps within the period $\Delta T = T_n - T_i$ there is an accumulation of risks causing imbalance of the compliance of communicative competences models with labor functions, which leads to violation of the LHC design operation mode.

The total impacts on personnel and infrastructure elements through the development and implementation of preventive measures F are presented by the lower graph in figure 2. After superposition of the specified graphs of emergency risk, the wavy dotted line F is obtained in figure 2, the position of which should not be above the line of acceptable risk Rn.

The upper boundary of the zone of acceptable risk is proposed to be considered as the boundary of the balance area of communicative competencies and regulated by professional standards of labor functions.

The results of the implementation of the developed mechanism for managing the development of personnel professional competencies of the loading and haulage company are presented in figure 3.

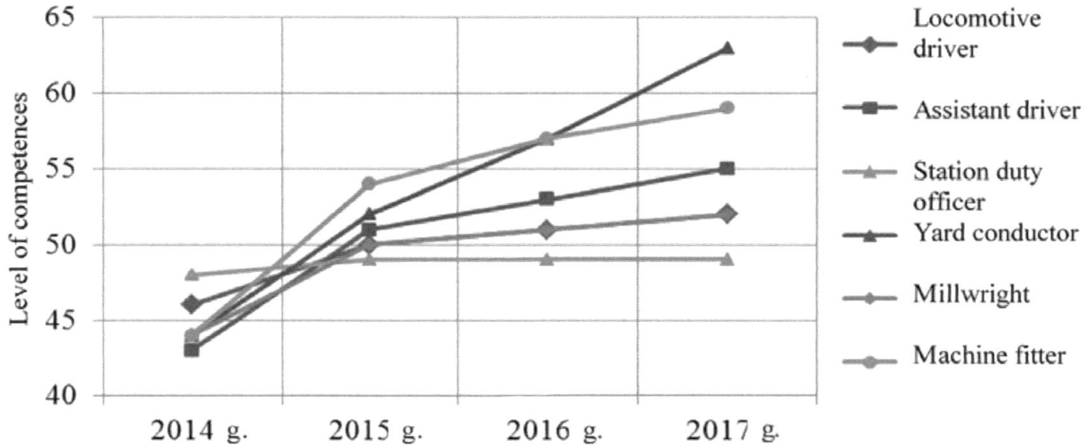

Figure 3. Dynamics of the professional competences of the personnel at "United Production and Transport Department of Kuzbass" LLC.

The graphs in figure 3 were obtained from the test results according to a 100-point assessment of employees. For the survey the international system of independent assessment of individual was used to select and develop the personnel of Hoqan Safety Assessment. As it can be seen from the graphs, during the last three years for almost all professions there is a rise in the level of competencies by 1.3-1.5 times, which confirms the effectiveness of the developed management mechanism for development of personnel professional competences.

3. Conclusion

As a result of solving the problem of managing the development of personnel professional competences, a mechanism was developed that minimizes the risks of emergency situations and accidents in the context of interaction between the workers of a loading and haulage department and production infrastructure elements. When implementing preventive measures that ensure that the risk is reduced to an acceptable one, it is proposed to take into account the influences of deviations in the

IOP Conf. Series: Materials Science and Engineering **354** (2018) 012013 doi:10.1088/1757-899X/354/1/012013

actual operating modes of the infrastructure elements from the normative ones regulated by labor functions.

References

[1] Professional standard: 27.092 *Specialist in Operation of Railway Transport in the Mining and Metallurgical Industry* approved on 01.02.2017, No. 121n
[2] Professional standard: 593 *Locksmith on Inspection and Repair of Rolling Stock of Railway Transport* approved on 02.12.2015, No. 954n
[3] Pristupa Yu D, Pavlova L D and V N Fryanov 2014 *Modeling the Process of Managing the Transport Flows of a Coal Holding under Uncertainty* (Moscow: Nauchnoye Obozrenie) p 216
[4] Pristupa Yu D, Shishkina S V et al 2015 *Proc. of the Intern. Sci. and Practical Conf in Knowledge based Tech. for the Development and Use of Resources* (Novokuznetsk: SibSIU) pp 195–200
[5] Shishkina S V et al 2015 *Proc. Xth All-Russian Conf. on Automation Systems in Education, Science and Production* (Novokuznetsk: SibSIU) 2015 166–173
[6] Shishkina S V, Pristupa Yu D, Pavlova L D and Fryanov V N 2016 *Proc. IVth All-Russian Conf. on Modeling and Knowledge-based IT in Tech. and Socio-economic Systems* (Novokuznetsk: SibSIU) pp 8–14
[7] Pristupa Yu D et al 2014 *Coal* **12** 50–52
[8] *RD 03-357-00 Methodological Recommendations on Drawing up a Declaration of Industrial Safety of a Hazardous Production Facility* approved by the resolution of the Gosgortekhnadzor of Russia on 26.04.2000 N 23

IOP Conf. Series: Materials Science and Engineering **354** (2018) 012014 doi:10.1088/1757-899X/354/1/012014

Technical devices of powered roof support for the top coal caving as automation objects

M S Nikitenko, S A Kizilov, P I Nikolaev and I S Kuznetsov

Federal Research Center of Coal and Coal Chemistry, Siberian Branch of the Russian Academy of Sciences, 18 Sovetskiy Ave., Kemerovo, 650000, Russia
T.F. Gorbachev Kuzbass State Technical University, 28 Vesennyaya Street, Kemerovo, 650026, Russia

E-mail: schum24@rambler.ru

Abstract. In the paper technical devices for the top coal caving as automation objects in the composition of the longwall mining complex (LTCC) are considered. The proposed concept for automation of the top coal caving process allows caving efficiency to be ensured, coal dilution to be prevented, conveyor overloading to be prevented, the shearer service personnel to be unloaded, the influence of the "human factor" to be reduced.

1. Introduction

The effective operation of the longwall complex depends on the coordinated work of the powered roof support, the front armored face conveyor (AFC) and the shearer. Therefore, as usual, integrated automation systems are used that provide electro-hydraulic control of the support from the control posts located on the support section and remote from the central control panel installed on the drift, with reference to the conveyor and shearer position.

Technology development of thick coal seams with controlled top coal caving involves the use of crepe sections with devices of adjustable coal withdrawal on the front AFC. The distinctive features of such supports are the availability of an outlet window, a feeder, which are intended for withdrawal and discharged coal loading on the front AFC [1, 2]. The feeder evenly releases the rock mass over the whole area of the outlet window. Its performance is adjustable in a wide range. It is possible to simultaneously operate a group of feeders on the front AFC. The number of feeders working in the group is determined by its productivity. If we assume that the feeders operate under the same conditions, their weighted average productivity can be adjusted to the required value of the full load of the conveyor when the feeder group operates [2].

Depending on the functions performed, the technical devices providing caved coal drawing on the front AFC are:

- backstop, providing overlapping withdrawal flow gate (outlet window) powered support section;
- feeder initiating the process of motion caved rock mass drawing;
- feeder tray, providing the function of loading on the front AFC with simultaneous operation of the feeder.

It is possible to implement a feeder and feeder tray in the form of a single device, depending on the type of feeder.

Content from this work may be used under the terms of the Creative Commons Attribution 3.0 licence. Any further distribution of this work must maintain attribution to the author(s) and the title of the work, journal citation and DOI.
Published under licence by IOP Publishing Ltd

Technical devices providing caved coal drawing are separate devices and at the same time are assembly units of a single powered support unit and functionally should be considered as an integral part of the longwall complex. For this reason, the development of the principles of automated control of the technical means of caved coal drawing on the front AFC, even at the initial level, should be considered as the development of a control subsystem, eventually integrated into the used longwall complex control system.

2. Methods of research

To substantiate underground automated geotechnologies with elements of robotization, without the constant presence of people in the faces, a system-functional approach in the form of IDEF0 methodology [3, 4] was applied. To this end, the top coal caving process is decomposed into technological operations and an IDEF0 model of the process of mechanized top coal caving (figure 1) is developed, displaying its organization without using automation tools.

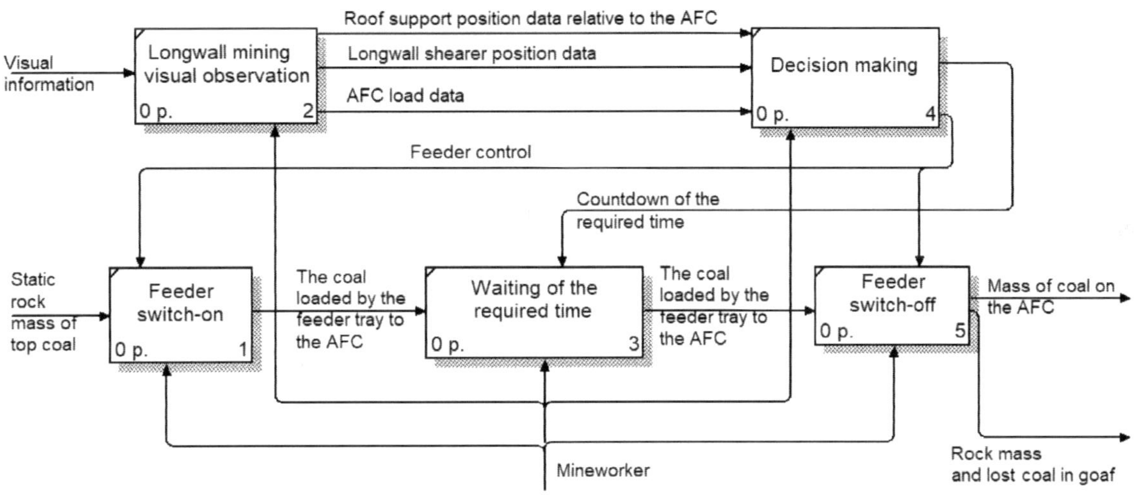

Figure 1. IDEF0-model of the mechanized top coal caving.

As we are visually modeled in figure 1, with mechanized caving, all technological operations are closed on the mineworker: he tracks the course of the coal-face work, the longwall shearer position and the roof support position, starts and stops the runner at the discharging time. When automation of the caving process, these functional tasks are transferred to the control system.

As a result of the analytical review of the existing control systems of longwall complexes, it was concluded that the automation of technical systems in the development of coal seams longwall complexes developed at a high level and is built, usually based on electrohydraulic control systems; The available technical solutions for the automation of technical devices for LTCC from thick seams to the rear conveyor [5, 6] are integrated through the provided free (reserve) control channels of the hydraulic cylinders powered support unit based on the produced automated electrohydraulic control systems.

The analysis of dissertations on the specialty "Robots, mechatronics and robotic systems" over the last 5 years has led to the conclusion that the trend of development of Russian robotics is aimed at the development of adaptive control systems for robots and mechatronic systems in nondeterministic conditions. Namely, to create sets of mechatronic and robotic systems aimed at solving a single task of control taking into account a changing external environment and target performance and reliability criteria due to the intellectualization of control systems.

If we consider technical devices for top coal caving withdrawal as objects of automation and robotics, then the mine conditions are quite deterministic in terms of technological and operational environment and possible operating modes. It does not require the constant adaptation of technical devices to the work environment. Allows the task of adaptive automation exclusively within the technological function to be considered, which is the determination of the moment of the beginning of top coal withdrawal (discharging) and the beginning of the appearance of rock fraction (fragments) in the withdrawal flow gate of the n-th support unit to ensure effective implementation of undulated areal extraction of coal.

3. Results and discussion

Undulated areal extraction involves the simultaneous opening and loading of coal from the group powered support units. Its effectiveness depends on the number and sequence of simultaneously open / closed release windows support units, whose work should be linked to the loading and position of the front AFC, as well as the position of the longwall shearer.

The duration of top coal caving withdrawal is proposed to be determined in real time at the time the rock fraction is exceeded in the discharging mass. For this, it is necessary to identify the different physical properties of coal and rock, which can be fixed during their discharging in the flow:

1. Different loose density of coal (200 kg/m^3) and enclosing rocks (over 300 kg/m^3). It is possible to distinguish between coal and rock by using the weight of rock mass established before the flow gate;

2. The ash content of coal affects its ability to reflect and pass through itself ionizing radiation and X-rays. Apparatus using this pattern, ash gauges, allows using its irradiation of a coal stream to detect its ash content by X-ray or gamma rays [7], which allows to stop coal withdrawal in time when the rock hits the flow gate of the powered support unit.

To ensure caving efficiency in the determination of ash content, it is proposed to install not one sensor, but several, higher than the first one. This will allow controlling the approach of a large rock mass and closing the flow gate until it falls onto the conveyor. Also, a system of several ash gauges will reduce the likelihood of premature closing of the outlet window and stopping the feeder when a single rock fragment passes through the working zone of the ash dump (figure 2).

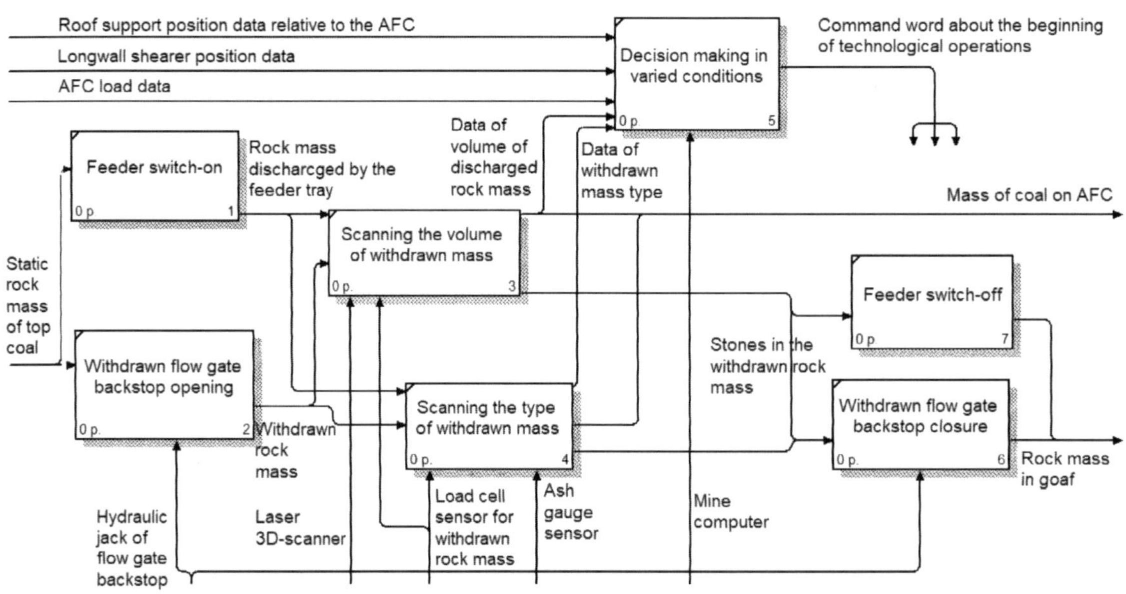

Figure 2. IDEF0-model of the automated top coal caving.

Thus, the control system for technical devices for top coal caving should provide the solution of the following tasks:

- timely opening / closing the outlet window n-th powered support unit to ensure the operation of the complex in a given mode (Undulated areal extraction);
- adaptive control of the backstop and feeder, depending on the composition of the withdrawn rock mass, congestion and the position of the conveyor.

Based on this, the concept of automated top coal caving with robotic elements, presented on the IDEF0 model in figure 2.

4. Conclusions

The proposed concept for automation of the top coal caving process makes it possible:

- ensure caving efficiency of top coal;
- prevent coal dilution when installing the coal-rock sensor above the outlet powered support unit to ensure that the backstop can be closed in advance;
- dosing output depending on the position and loading of the conveyor, shearer position;
- prevent conveyor overloading;
- unload the shearer service personnel, who are directly in the face;
- reduce the influence of the "human factor" in managing the process of top coal caving.

On the basis of the IDEF0 model obtained, to conduct a detailed study of the structural and functional schemes of the adaptive system with the robotization element aimed at solving the problem of controlling technical devices for top coal caving taking into account the changing parameters of the technological environment.

Acknowledgements

The project was financially supported by the Ministry of Education and Science of the Russian Federation within the Federal target program 'Research and development of priority directions of scientific-technological complex development of Russia for 2014-2020' (agreement No. 14.604.21.0173 from 26.09.2017, unique ID RFMEF160417X0173).

References

[1] Klishin V I, Anferov B A and Kuznetsova L V 2017 *Proc. of Int. Sci. and Practical Conf. in Innovations in the Fuel and Energy Complex and Engineering* pp 57–63
[2] Klishin V I and Klishin S V 2010 *Physical and Technical Problems of Mining* **2** 68–80
[3] Shirobokova S N 2014 *Use of Tools to Support Business Process Reengineering* (Novocherkassk: Platov South-Russian State Polytechnic University) p 194
[4] Nikolaev P I and Zinoviev V V 2016 *Bulletin of Kuzbass State Technical University* **4** 26–33
[5] *Customized Roof Support Systems* www.cat.com/en_US/products/new/equipment/ underground-longwall/roof-supports/18346354.html
[6] *TIANDI Science&Technology Co.* www.tdtec.com
[7] *Physical Methods for Controlling the Ash Content of Coal* 1978 (Moscow: Nedra) pp 53–56

IOP Conf. Series: Materials Science and Engineering **354** (2018) 012015 doi:10.1088/1757-899X/354/1/012015

Longwall top coal caving (LTCC) mining technologies with roof softening by hydraulic fracturing method

V Klishin, S Nikitenko and G Opruk

Federal Research Center of Coal and Coal Chemistry, Siberian Branch of the Russian Academy of Sciences, 10 pr. Leningradksii, Kemerovo, 650065, Russia

E-mail: nsm.nis@mail.ru

Abstract. The paper discusses advanced top coal caving technologies for thick coal seams and addresses some issues of incomplete coal extraction, which can result in the environmental damage, landscape change, air and water pollution and endogenous fires. The authors put forward a fundamentally new, having no equivalent and ecology-friendly method to difficult-to-cave roof coal – directional hydraulic fracturing and nonexplosive disintegration.

1. Introduction

Most countries in the world produce coal using high-output fully mechanized longwall mining technology (round 84% of total coal production). Sustainable efficiency of such mines requires maintaining all-time high output per face (more than 4 thousand tons per day), which is only possible in favourable ground conditions [1]. The effective range of the fully mechanized longwall systems can be expanded by means of cutting difficult-to-cave roof strata. Under such conditions, unexpected and uncontrollable rock falls are highly probable, which is extremely dangerous to mine personnel, mining equipment and roadways. Furthermore, overhanging roof induces concentrated rock pressure in production face area and its junctures with roadways, which can trigger coal and gas outbursts.

In rockburst-hazardous and outburst-hazardous coal mining, overhanging difficult-to-cave roof increases stresses on coal face edges, which provokes dynamic and gas-dynamic events. At the moment of roof caving, gas releases from fallen roof rocks in roadways, which violently increases methane content of mine air. Gas bursts are accompanied with powerful shock waves and can induce rock failure or inflammation and explosion of gas and coal dust.

Many researchers address issues connected with thick coal strata mining.

Mondal D., Roy P.N.S. and Behera P.K. (Coal India Limited, Indian School of Mines) in the article "Use of correlation fractal dimension signatures for understanding the overlying strata dynamics in longwall coal mines" prove that longwall mining is considered to be the best coal mining practice due to vast recovery of coal over other forms of underground as well as opencast mining methods.

Duncan G., Sobey G. and Clarke T. (Austar Coal, DBT Australia) in their study "Top coal caving longwall maximizes thick seam recovery: Austar's longwall system offers opportunities in seams thicker than 4.5 meters" share hands-on experience on the Longwall Top Coal Caving (LTCC) method used by Austar mine, owned and operated by Yancoal Australia, for extracting coal seams thicker than 4.5 m.

In the work "Three-dimensional experimental study of loose top-coal drawing law for longwall top-coal caving mining technology," Jiachen Wang, Jinwang Zhang, Zhengyang Song and Zhaolong Li (China University of Mining and Technology) investigate the loose top-coal drawing

Content from this work may be used under the terms of the Creative Commons Attribution 3.0 licence. Any further distribution of this work must maintain attribution to the author(s) and the title of the work, journal citation and DOI.
Published under licence by IOP Publishing Ltd

law of the longwall top coal caving (LTCC) mining technology using self-developed 3D test device based on the loose medium flow field theory.

According to the analysis of the coal resource base of the Russian Federation, in-place reserves of coal mines contain 1/3 of unfavorable and 1/10 of extremely unfavorable coal unsuitable for fully mechanized longwall mining technology. By expert estimates, not more than 1/3 of known coal reserves are suitable for fully mechanized longwall mining technology, which is an impeding factor to competitive ability of the coal mining industry in the world market, and 2/3 of high-quality coal occurs in steeply dipping seams. The difficult ground conditions are one of the sources of problems in the present-day mining. Thus, efficiency and ecological friendliness of coal mining is the top-priority objective in Russia, and the prospects of coal mining in the country depend on the success of reaching that objective.

2. Methods

Not all underground coal mining technologies are efficient and ecology-friendly. At the stage of mine planning and design, coal recovery rate is calculated based on the value of project coal loss, including irrecoverable loss of coal within mine fields. At the same time, coal recovery rate is in many ways governed by ground conditions, specifically, by thickness and dip angle of coal seams, by depth of mining and by mining technology. Incomplete extraction of coal may result in endogenous fires in mined-out areas, which is hazardous to health and tangibles. Landscape changes, air and water pollution, methane emission and other negative factors take place at the same time.

Coal mine safety rules state that mining of spontaneously inflammable coal should involve full-scale backfilling. Furthermore, in case of thick inflammable coal seams, it is required to make separate extraction panels with fire barriers in-between, which reduces coal recovery rate.

At this conjuncture, it is required to come up with an efficient process solution to ensure competitive ability of mines cutting thick coal seams.

The international practice of thick coal seam mining involves multi-slice working and one-slice longwall mining technologies. Multi-slice working causes high coal loss and, as a consequence, induces risk of endogenous fires. For this reason, the most preferred method is one-slice thick coal seam mining with top coal or interlayer coal caving [1].

There are two known scenarios of mechanizing longwall top (interlayer) coal caving method – using a front face conveyor, e.g. KTU and KNKM (manufactured in Russia) or VHP-731 (Hungary) and using a complete longwall armored rear face conveyor, e.g. OKPV-70 and KM81V system (Russia), or AFC (China). The latter variant has widely been introduced in China, where there are many thick and super thick coal seams and where high-productive coal mines reach annual coal output of 4.5–8.00 million tons with one or two operating longwalls. Different types of powered roof support connected with the features of longwall mining technologies have been successfully trialed in Yangzhou, Yangquan, Luan and Datong [2, 3].

The advantages of top (interlayer) coal caving with the complete longwall armored face conveyor include appreciable reduction in preparatory works, in capital and operating costs, power consumption and in spontaneous inflammation of coal; moreover, it allows coal mining in difficult ground conditions and enables extracting coal left in pillars. All these benefits enhance efficiency and safety of mining at the increased output per face and higher mining rate.

The new technology proposed for thick coal seams is implemented with the help of mechanized systems with robotic coal discharge using feeders with adjustable capacity [4, 5]. The design of the powered roof support with the controllable coal discharge to face conveyor takes into account geomechanical processes running in coal and roof rocks, involves advantages of the similar known designs and eliminates their shortcomings. This offers a new vision of advancement in the science-based technologies of thick coal seams longwall mining

The proposed technology enables one-slice cutting of coal seam 8.0–10.0 m thick at the coal recovery rate of not less than 0.8 and the projected output per face up to 15 thousand tons per day.

A similar approach is intended to be used in underground mining of thick steeply dipping coal seams with sublevel caving method.

The domestic industry currently does not manufacture high-capacity longwall systems suitable for efficient and safe mining of thick steeply dipping coal seams. Thus, thick steeply dipping coal is not extracted using underground mining methods because of the lack of appropriate mining machines.

One more competitive advantage of the recommended technology is its applicability in thick steeply and super steeply dipping coal seams using travelling powered shield support.

The proposed technologies avoid drilling-and-blasting but involve a new, having no equivalent approach to difficult-to-cave roof—method of directional hydraulic fracturing.

The currently available techniques of softening difficult-to-cave roof coal (advance torpedoing, hydromicrotorpedoing, etc.), despite the many years-long experience, are not always efficient for a few common drawbacks – non-uniformity and uncontrollability of rock breakage. The proposed method of directional hydraulic fracturing (DHF) integrates lamination and separation of rocks per blocks, which avoids roof overhanging and eliminates abrupt dynamic events in the time of immediate and subsequent bents, allows more uniform change in mechanical properties of rocks and ensures safety in reused roadways. A dedicated equipment set has been designed and tested in mines. Based on the multiple experimentation on application of DHF to softening roof rocks in assembling rooms in mines, as well as on the ground of visual observations, figure 1 shows a schematic pattern of roof caving when the fully mechanized longwall system comes out of an assembling room.

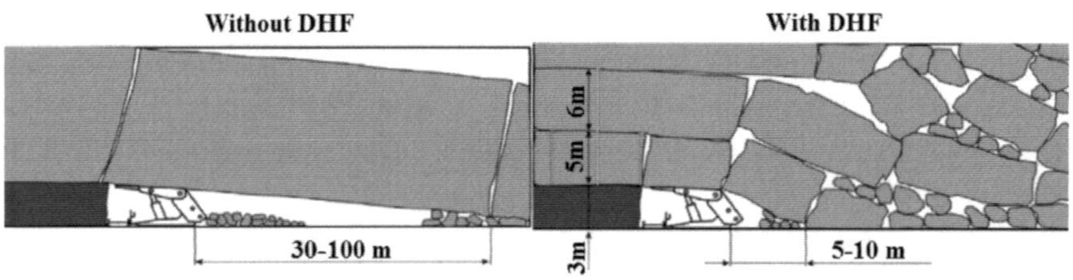

Figure 1. Cutting of difficult-to-cave roof with and without softening by DHF.

3. Results and discussion

The full-scale practice shows that lamination of hard solid roof rocks not only offers comfortable working conditions for powered support and cutter–loaders but enables preventing such dynamic events as rock bursts and coal and gas outbursts. Currently, the described approach is widely introduced in Kuzbass mines in Russia [6] and is included by Polish practitioners in Pokoj Mine in various process flowsheets as the basic procedure to prevent rock bursts [7].

Pre-mining methane drainage faces some technical difficulties as there is no developed network of drainage surfaces and channels before mining is started. The available degassing techniques allow recovery from 20 to 30% of total methane released. Because of this inefficiency of the current drainage technologies, many mines operating in coal seams featuring high gas content in Russia set a gas content limit, which hinders reaching high mining rate and high output per face. Thus, there is a paradoxical situation in the coal mining industry: the technical capability is available but unimplementable. This is also the explanation of low efficiency of labor and high injury rate. In the meanwhile, the modern strategy of underground coal mining anticipates high concentration and intensification of mining.

Aimed to enhance efficiency of pre-mining gas drainage in unloaded coal seams and for current degassing of coal after pressure relief, a fundamentally new approach to reaching maximum well rate has been developed—multiple directional hydraulic fracturing in a degassing hole (figure 2) [8].

The method consists in creation of long highly permeable drainage channels in the vicinity of a hole by means of opening of natural joints when new fractures grow. In this case, the cumulative methane content reduction in mine air in the operating zone of a cutter–loader is higher than in the conventional degassing method, and the implementation period is 3–4 times shorter at the complementary effect of decreased dust content.

To this effect, the process flowsheets are developed and the full-scale equipment system of DHF has been designed and tested in mines.

The accomplished research and development make the framework for the technology that ensures reduction in the degassing drilling by 3 times and more at the simultaneous increase in the rate and depth of drainage not less than by 2 times at the final air-and-methane mix recovery. As a result, safety of coal mining is enhanced due to pre-drainage, risk of methane outbursts and rockburst hazard is mitigated owing to partial stress relief in rock mass due to fracturing, and harmful methane emission is reduced.

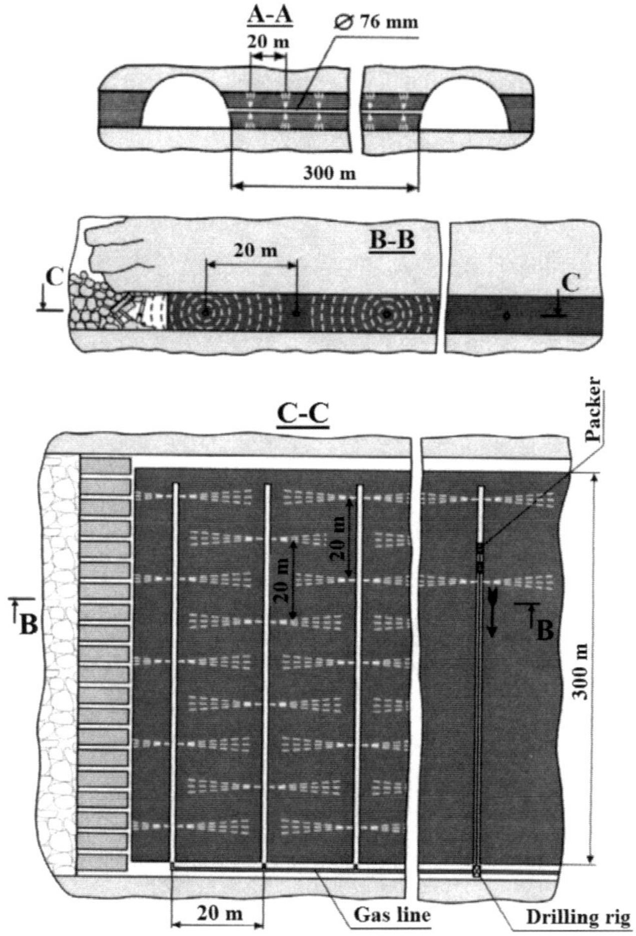

Figure 2. Multiple hydraulic fracturing pattern.

4. Conclusions

One of the key influences on the efficiency of the described longwall mining system and technology are the thickness of coal seams and breakability of coal under effect of rock pressure. The present paper authors put forward a thick top coal caving technology using the modern powered roof support with the controllable coal discharge. The fully mechanized longwall system design involves feeders arranged in the powered support units to ensure adjustable coal discharge along the whole length of the longwall, which enables enhanced economic performance of longwalling.

Parameters of powered roof support are inter-dependent and governed by process steps of advance of the powered support units; a rational travel step is selected with regard to the width of discharged coal flow, divisible by the total number of the powered roof support advance steps, or to the width of longwall. The both designs of powered roof support for cutting thick coal seams with the controllable discharge to front or rear armored face conveyor offer new prospects for thick coal seam mining using underground methods. The potential advantages include:

- one-slice longwalling, which promotes high concentration of mining, reduces production loss and abates risk of endogenous fires;
- reduction in volume and cost of development heading and maintenance by 1.5–2.0 times;
- utilization of rock pressure and cavability of top coal, which cuts down coal production cost;
- decrease in cost of longwall equipment and coal haulage within a mine field;
- decline in value of assembling and disassembling of longwall systems, haulage machines, electric equipment, pipelines, preventive control, etc;
- saving of cost of produced coal at the concurrent sharp increase in output per face given relatively slight growth in manpower.

The proposed technologies are meant for the coal producing regions in Russia, India, Georgia and other countries engaged in mining of thick coal seams. On the other hands, subject to relevant modification, these technologies are adaptable to thick salt rock beds and diamond-bearing strata.

Acknowledgements
The study has been supported by the Ministry of Education and Science of the Russian Federation in the framework of the Federal Targeted Program on R&D in Priority Areas of Advancement in the Science and Technology of Russia for 2014–2020, Topic: Development of Efficient Robotic Longwall Top Coal Caving Technology and Equipment (Unique Identifier RFMEF160417X0173).

References

[1] Klishin S V, Klishin V I and Opruk G Y 2013 *Journal of Mining Science* **49** 932–940
[2] Jinshuai Guo, Liqiang Ma, Ye Wang and Fangtian Wang 2017 *Energies* **10** 1371
[3] Melnik V V and Suchchev R A 2009 *Mining Information and Analytical Bulletin – GIAB* **5** 198–210
[4] Nikitenko M S and Malakhov Yu V 2017 *High Technologies of Mineral Mining and Use: Scientific Journal* **3** 484
[5] Kizilov S A, Neogi B, Nikitenko M S, Nikolaev P I and Kuznetsov I S 2017 *Mining Industry Journal* **6(136)** 96-99
[6] Klishin V I, Buchatsky V M and Konovalov L M 2007 *Coal* **6** 40–43
[7] Jevecki Y 2002 *Gluckauf* **2(3)**
[8] Klishin V I, Kokoulin D I, Kubanychbek B and Durnin M K 2010 *Coal* **4**

IOP Conf. Series: Materials Science and Engineering **354** (2018) 012016 doi:10.1088/1757-899X/354/1/012016

Concept of mobile operator position based on neurocomputer interface and augmented reality

S A Kizilov[1,2], M S Nikitenko[1,2] and B Neogi[3]

[1]Federal Research Center of Coal and Coal Chemistry, Siberian Branch of the Russian Academy of Sciences, 18 Sovetskiy Ave., Kemerovo, 650000, Russia
[2]T.F. Gorbachev Kuzbass State Technical University, 28 Vesennyaya Street, Kemerovo, 650026, Russia
[3]CoE & CII Centre for Innovation , JIS College of Engineering (JIS Group India), Block A, Phase III, Dist. Nadia, Kalyani, West Bengal, 741235, India

E-mail: schum24@rambler.ru

Abstract. The concept of creating a mobile operator position for the mining industry based on combining the technologies of the neurocomputer interface and augmented reality is proposed.

1. Introduction

The readily available and rich mineral deposit of the mining complex in Russia and abroad is now almost exhausted. Reserves that are characterized by complex mining and geological conditions are involved in development - stratification depth increases, occurrence is noted in steep thick seams, high gas content of coal seams. However, the mining equipment used and modern mineral extraction technologies from thick and steeply inclined thick coal seams are not without the constant presence of people in the longwall area, which is due to the complexity of the mining technology. In this case, mistakes in mining machine control or the influence of the human factor on the quality of machine control of a more complex technological process, accordingly increases the risks of occurrence of emergency situations. Therefore, the further development of the mining industry and the improvement of the mining safety system depend to a greater extent not only on the use of new modern types of equipment and processing technologies. An important factor is the dynamics and quality of creation, as well as the depth of introduction of modern automated control systems for technological processes and dispatching of mining operations.

The need to increase the efficiency of the development of difficult-to-extract minerals in complex mining and geological conditions and in mines dangerous for gas and dust creates the need for a transition to the "manless face" extraction technology for automated and automatic equipment brought to the level of robotic complexes [1].

Based on the foregoing, not only the development of new technical solutions for working out thick and steeply inclined mineral deposits [2], but also modern automated controls and dispatching equipment of the mining complex is topical. Developed new technological development schemes [3] and technical equipment for extracting minerals [4, 5], automated control systems of mining equipment and its diagnostics [6 – 11] allow to form the basis of the concept of creating robotic complexes for mining and extraction of reservoir deposits of minerals (coal, diamond-bearing ore, etc.) with a control system that takes into account, including the state of the mountain massif and its dynamic manifestations. The first step to its creation is the automation of process control during the

Content from this work may be used under the terms of the Creative Commons Attribution 3.0 licence. Any further distribution of this work must maintain attribution to the author(s) and the title of the work, journal citation and DOI.
Published under licence by IOP Publishing Ltd

IOP Conf. Series: Materials Science and Engineering **354** (2018) 012016 doi:10.1088/1757-899X/354/1/012016

development of thick coal seams with top coal caving technology (LTCC), which should concern both the operational withdrawal and the control and dispatching system of the equipment of the longwall complex.

2. Methods of research

With regard to the control and dispatching system for the operation of the longwall complex equipment, when working out thick coal seams with top coal caving, we note that the classification of the causes of emergencies with the influence of the human factor into three main categories:

- accidental error, inattention or weakening of control over the situation;
- low qualification of maintenance staff;
- a gross violation of safety precautions, often combined with the first two points,

makes it possible to draw a conclusion that modern dispatching systems have two serious drawbacks: 1 – stationary place of the operator (means the presence of devices or monitors with indication of the current state of the system and a fixed location of controls); 2 – high requirements for the qualification and physical condition of the operator.

In this context, the hardware based on the Brain computer interface (BCI) is very promising and relevant for building an automated control system, in conjunction with augmented reality equipment

Currently, the authors, in conjunction with JIS Collage of Engineering, where specialists are already working on the study of basic BCI-based control modules [9, 10], are working on the concept of the mobile operator's position and integration of BCI equipment and augmented reality into the automated control system . In addition, it is possible to expand the management methods by using the electromyography method (EMG).

The main aim is to organize the work of the dispatcher of the automated control system for technological processes of underground mining of mineral deposits at a new technical level, without permanent location in the workplace, with a reduction in the probability of making an error, loss of visual contact with the parameters of the control object, based on virtualization of control and monitoring instruments.

The main idea is combining the technologies of the neurocomputer interface, electromyography and augmented reality for the control system of technological processes of underground mining of mineral deposits.

3. Results and discussion

The ability to display information about the state of the complex of equipment directly in front of the operator with the rendering of interactive controls of the complex is provided by the augmented reality systems, access to the main emergency management functions takes place through the adaptive automated control system based on the BCI control interface technology. The system based on BCI allows you to get an almost instantaneous signal from the operator and perform a specified action, which is important in an emergency situation. In this case, the operator does not even need to simulate pressing the "emergency stop button", the general scheme of the mobile workstation of the operator longwall complex is shown in figure 1.

The mobile operator's position of the operator allows solving a number of problems related to the safety of work in underground workings. Such problems as loss of eye contact for a long time with indicators that display information about the status of the working complex are being solved. There is an opportunity to better equip the operator's workplace ergonomics and reduce his fatigue. The operator is given the opportunity to make control manipulations at any point in the control room, and not only when he is directly at the stationary console.

The control of the equipment within the framework of the proposed concept takes place through the interaction of the operator's hands with virtual keys displayed on the display of the augmented reality headset, the operator's actions are monitored by two different systems, the data from which are compared and their verification takes place.

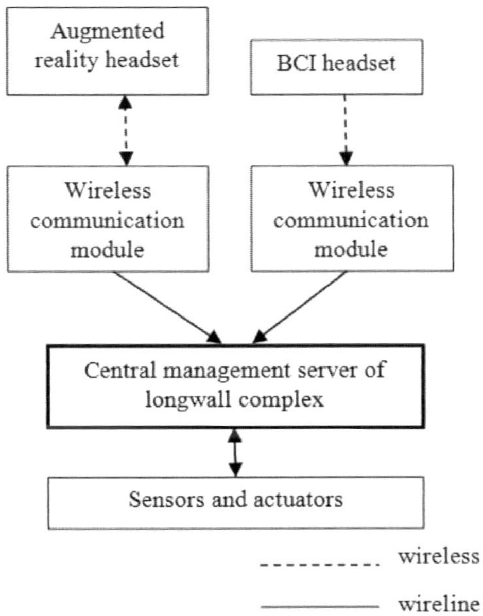

Figure 1. General scheme of the mobile workstation of the operator of the longwall mining complex.

The first system is a standard method for virtual and augmented reality systems, operating on the basis of information received from built-in video cameras and position sensors. But for high accuracy of operation, such a system requires the presence of markers on the monitored parts of the body, which are not very convenient in everyday use.

The second system is based on the use of methods of electromyography, the system tracks the impulses of the muscular system of the operator and translates them into commands for controlling the technique. Such a system, although it requires the mounting of sensors on the operator's body, but they should not be felt if installed properly.

It is worth noting that the technical means based on the neurocomputer interface, allowing the introduction of an additional control channel from the dispatcher or operator (user) of equipment, have recently entered the market, (since 2003, Mindball (Interactive Productline), MindSet, MindWave (NeuroSky), Neural Impulse Actuator (OCZ Technology), EPOC и Emotiv Insight (Emotiv Systems), Mindflex (Mattel), XWave headset (PLX Devices), MyndPlay BrainBand (MyndPlay), OpenBCI (OpenBCI project), Muse (InteraXon) etc. insufficiently researched and tested in real conditions in the direction of application in industrial enterprises for a solution for giving dispatching. In addition, the psychophysiological state of the operator has a significant influence on the quality and efficiency of generation of control actions.

4. Conclusion

The proposed concept of mobile operator position makes it possible to organize the work of the dispatcher of the automated control system for technological processes of underground mining of mineral deposits without permanent location in the workplace, with a reduction in the probability of making an error, loss of visual contact with the parameters of the control object.

Acknowledgement

The project was financially supported by the Ministry of Education and Science of the Russian Federation within the Federal target program 'Research and development of priority directions of scientific-technological complex development of Russia for 2014-2020' (agreement No. 14.604.21.0173 from 26.09.2017, unique ID RFMEF160417X0173).

References

[1] Ermekov T E, Nesipbaev Zh S and Sarsenbaev T U 2001 *Works of the University* **1** 46–47

[2] Malakhov Yu V 2017 *Proc. of the Int. Sci. and Practical Conf. in Innovations in the Fuel and Energy Complex and Engineering* pp 62–68

[3] Klishin V I 2013 *Mining Informational And Analytical Bulletin* **6** 36–47

[4] Klishin V I and Kokoulin D I 2015 *A Powered Suport for Working out Thick Steep Coal Seams with Sublevel-stoping* patent RF no. 1604472 U1

[5] Klishin V I, Anferov B A and Kuznetsova L V 2017 *Proc. of the Int. Sci. and Practical Conf. in Innovations in the Fuel and Energy Complex and Engineering* pp 57-63

[6] Roiter M, Wexler J and Kurfürst V 2007 *Industry of Kazakhstan* **9** 34–40

[7] Sadovets V Yu and Kizilov S A 2016 T*echnologies and Materials* **3** 4–7
[8] Nikitenko M S, Malakhov Yu V, Neogi B, Chakraborty P and Banerjee D 2017 *IOP Conference Series: Earth and Environmental Science* **84** 012002
[9] Nikitenko M S 2016 *IOP conference series: Earth and environmental science* **84** 012007
[10] Neogi B et al 2016 Cognitively Velocity Controlled Vehicle Patent IND File no. 201631017186 Indian Patent Journal no. 26/2016 http://www.ipindia.nic.in/writereaddata /Portal/IPO Journal/ 1_350_1/part1.pdf
[11] Neogi B et al 2016 *Thought Concentration Controlled Dexterous Prosthetic Arm for Handicapped.* File no. 201631017174. Indian Patent Journal no. 26/2016. http://www.ipindia.nic.in/writereaddata/Portal/IPOJournal/1_350_1/part1.pdf

IOP Conf. Series: Materials Science and Engineering **354** (2018) 012017 doi:10.1088/1757-899X/354/1/012017

Equivalent model of a dually-fed machine for electric drive control systems

I Yu Ostrovlyanchik and I Yu Popolzin

Siberian State Industrial University, 42 Kirova street, Novokuznetsk, 654007, Russia

E-mail: eidoline@yandex.ru

Abstract. The article shows that the mathematical model of a dually-fed machine is complicated because of the presence of a controlled voltage source in the rotor circuit. As a method of obtaining a mathematical model, the method of a generalized two-phase electric machine is applied and a rotating orthogonal coordinate system is chosen that is associated with the representing vector of a stator current. In the chosen coordinate system in the operator form the differential equations of electric equilibrium for the windings of the generalized machine (the Kirchhoff equation) are written together with the expression for the moment, which determines the electromechanical energy transformation in the machine. Equations are transformed so that they connect the currents of the windings, that determine the moment of the machine, and the voltages on these windings. The structural diagram of the machine is assigned to the written equations. Based on the written equations and accepted assumptions, expressions were obtained for the balancing the EMF of windings, and on the basis of these expressions an equivalent mathematical model of a dually-fed machine is proposed, convenient for use in electric drive control systems.

1. Introduction

A dually-fed machine (DFM) is generally referred to as a connection circuit of asynchronous motor with a phase rotor in which a voltage from the external source is applied to the rotor, with the possibility of adjusting the amplitude, frequency, and phase of this voltage [1]. The use of a DFM has significant advantages over the use of an asynchronous motor with a phase rotor and a rotor station, since it allows the flow of slip energy to be controlled and high energy efficiency to be achieved. To control the DFM in the electric drive control systems, it is necessary to build its mathematical model. The task of this paper is to obtain a simple DFM model suitable for the above purposes.

Mathematical description of the dually-fed machine is very complicated because the machine moment is a non-linear function of the winding currents of the stator and rotor, so its identification is difficult. Besides, due to the presence of additional voltage source in the rotor, methods of mathematical modeling that are well suited to describe the asynchronous motor with a voltage source only in the stator are inconvenient for describing the DFM.

2. Statement of the problem

In [1] and [2], a DFM simulation using the method of an orthogonal coordinate system oriented along the representing vectors of machine coordinate was proposed. To simplify the mathematical model the authors chose a coordinate system associated with the representing vector of the stator current. In such a coordinate system, the model of electrical and electromechanical transformations in the DFM is written as a system of equations (1).

Content from this work may be used under the terms of the Creative Commons Attribution 3.0 licence. Any further distribution of this work must maintain attribution to the author(s) and the title of the work, journal citation and DOI.
Published under licence by IOP Publishing Ltd

IOP Conf. Series: Materials Science and Engineering **354** (2018) 012017 doi:10.1088/1757-899X/354/1/012017

$$\begin{cases} i_{Sg} = \dfrac{1/R_S}{T_S p+1}\left(u_{Sg} + pL_\mu i_{Rg} + \omega_g L_\mu i_{Ri}\right), \\ i_{Rg} = \dfrac{1/R_R}{T_R p+1}\left(u_{Rg} - pL_\mu i_{Sg} + \omega_g s L_R i_{Ri}\right), \\ i_{Ri} = \dfrac{1/R_R}{T_R p+1}\left(u_{Ri} + \omega_g s L_R i_{Rg} - \omega_g s L_\mu i_{Sg}\right), \\ M = \dfrac{3}{2} z_P L_\mu i_{Sg} i_{Ri}, \\ M - M_C = J_\Sigma \dfrac{d\omega}{dt}, \end{cases} \quad (1)$$

where

i_{sg} – projection of the representing vector of stator current on the axis g of the rotating orthogonal coordinate system g-i,

i_{Rg}, i_{Ri} – projections of the representing vector of rotor current on the axis g and i of the rotating orthogonal coordinate system g-i, respectively;

u_{Sg} – projection of the representing vector of stator voltage on the axis g of the rotating orthogonal coordinate system g-i, respectively;

u_{Rg}, u_{Ri} – projections of the representing vector of stator voltage on the axis g and i of the rotating orthogonal coordinate system g-i, respectively;

T_S, T_R – electromagnetic constants of stator and rotor time, respectively;

R_S, R_R – active resistance of the stator and rotor, respectively;

L_μ – magnetizing inductance of the machine;

z_P – number of pole pairs of the machine;

M, M_c – electromagnetic moment of the machine and the moment of loading, respectively;

J_Σ – total moment of inertia of the working mechanism brought to the rotor of the machine;

ω – angular speed of rotation of the machine rotor;

ω_g – angular velocity of rotation of the coordinate system, for the selected system determined by the expression $\omega_g = \dfrac{\arccos\left(\dfrac{u_{Sg}}{u_{S\max}}\right)}{T} = f_c \cdot \arccos\left(\dfrac{u_{Sg}}{u_{S\max}}\right) = f_{\omega_g}\left(u_{Sg}\right)$, where fc – the voltage frequency on the stator, $u_{S\max}$ – the voltage amplitude on the stator;

s – relative frequency of the emf of the rotor, determined by the expression $s = \dfrac{\omega_g - \omega}{\omega_g} = \dfrac{E_S - E_R}{E_S}$, where $E_S = k_E \omega_g$ is the emf of the stator, $E_R = k_E \omega$ – emf of the rotor,

k_E – coefficient of internal feedback on the motor emf;

p – differentiation operator.

The system of equations (1), taking into account the indicated notations, can be associated with the DFM structural diagram, shown in figure 1.

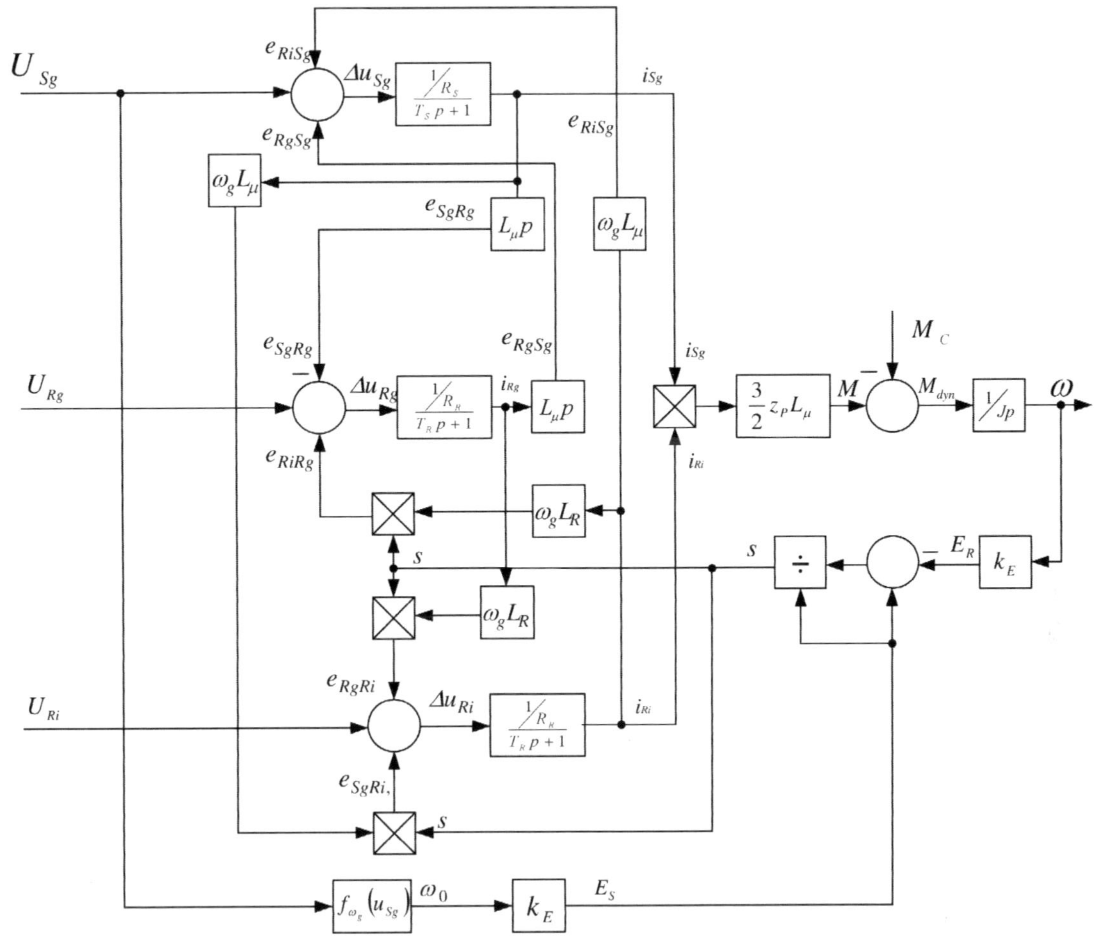

Figure 1. Structural diagram of the dually-fed machine in asynchronous mode in the orthogonal coordinate system associated with the stator current representing vector.

The DFM structural diagram shown in figure 1, although has simple transfer functions for windings, has a drawback – the presence of cross feedbacks, the corresponding emf, induced by the windings of opposite phases of the stator and rotor. In [1] it is proved that the equations of electric equilibrium of induction motor along the axes of coordinate system can be written in the operator form in the following form:

$$u = \frac{1}{k_e}(T_e p + 1)\cdot y + f_C\left(\bar{i}_S, \bar{i}_R, \bar{\psi}_S, \bar{\psi}_R, f_R\right), \qquad (2)$$

where u – projection of the generalized vector of stator or rotor voltage on the corresponding axis of orthogonal coordinate system,
T_e – equivalent time constant of the machine;
k_e – equivalent rate of gain of the machine;
i – projection of the generalized stator or rotor current vector on the corresponding axis of the orthogonal coordinate system;
$f_C\left(\bar{i}_S, \bar{i}_R, \bar{\psi}_S, \bar{\psi}_R, f_R\right)$ – compensation function of mutual influence of phase windings of the stator and rotor, depending on the selected coordinate system.

Provided that the stator and rotor phases of the machine are symmetrical and the currents and voltages in the windings are sinusoidal, we can assume that these emfs are small in magnitude compared to the emf induced by the windings stator phases in the windings of the same phases of the rotor and vice versa, and are balanced by the emf of self-induction, appearing in the respective windings. The compensation function in this case is determined by balancing emfs, which makes it possible to simplify the proposed mathematical model of the DFM and to obtain a simpler equivalent model of the machine.

Balancing emf in the chosen coordinate system in accordance with the above considerations, the equations of the system (1) and the scheme shown in figure 1, is determined by the expressions (3):

$$\begin{gathered}\Sigma e_{RgSg} = L_\mu \frac{di_{Rg}}{dt}, \Sigma e_{RiSg} = L_\mu \omega_g i_{Ri}, \Sigma e_{SgRg} = L_\mu \frac{di_S}{dt}, \\ \Sigma e_{RiRg} = \omega_g s L_R i_{Ri}, \Sigma e_{RgRi} = \omega_g s L_R i_{Rg}, \Sigma e_{SgRi} = \omega_g s L_\mu i_{Sg},\end{gathered} \tag{3}$$

where Σe is the balancing emf;
index S – stator winding;
index R – winding of the rotor;
index g – winding on the axis g of the two-phase DFM model;
index i – winding on the i-axis of the two-phase DFM model.
The DFM structural scheme, taking into account the balancing emf (3), is shown in figure 2.

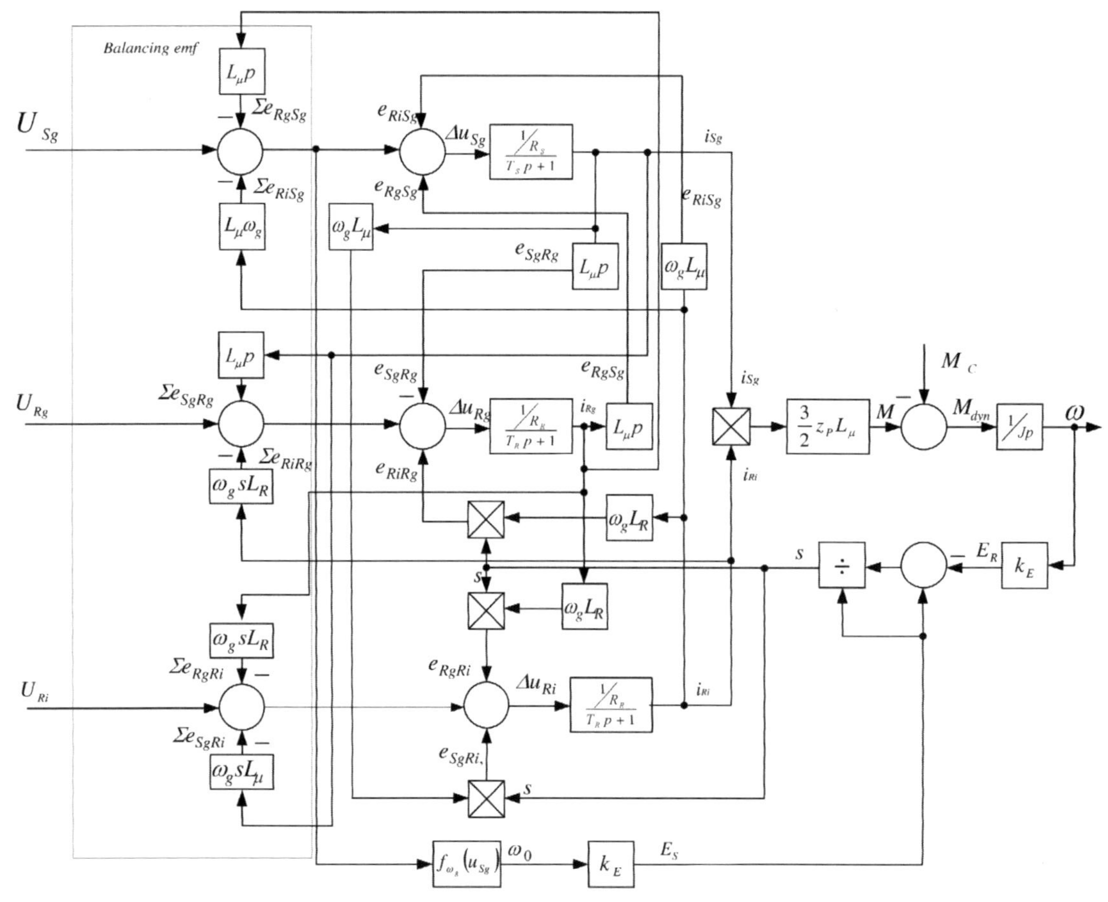

Figure 2. DFM structural diagram with the balancing of emf.

Taking into account the accepted assumptions, expressions (3) and signs of emf, the following relations are valid for the balancing ones:

$$\begin{aligned}&\Sigma e_{RgSg} \approx -e_{RgSg}, \Sigma e_{RiSg} \approx -e_{RiSg}, \Sigma e_{SgRg} \approx -e_{SgRg},\\&\Sigma e_{RiRg} \approx -e_{RiRg}, \Sigma e_{RgRi} \approx -e_{RgRi}, \Sigma e_{SgRi} \approx -e_{SgRi}.\end{aligned} \quad (4)$$

In the structure diagrams shown in figures 1 and 2, the controlling signals (forming the moment of the machine) are u_{Sg} and u_{Ri}. In the DFM, as a rule, the voltage at the stator u_S, and, consequently, its projection u_{Sg} are constant [1]. Let us consider the diagram shown in figure 2 with respect to the control action u_{Ri} under constant control action u_{Sg}. The DFM structure diagram, shown in figure 2, with the constant voltage at the stator and observing condition (4) takes the form shown in figure 3.

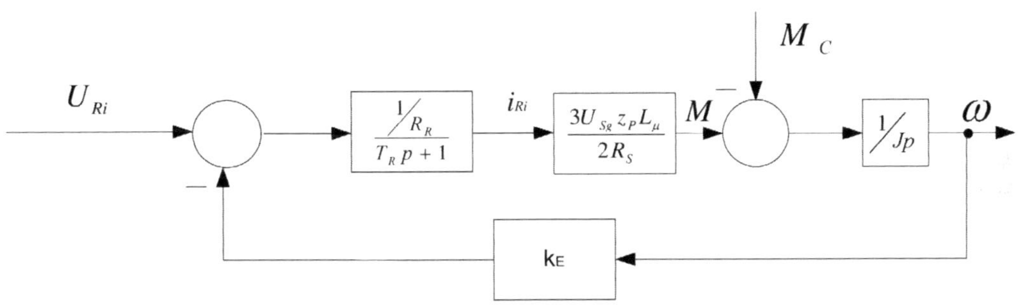

Figure 3. Equivalent structure diagram of a dually-fed machine.

$$k_E = L_\mu I_S = L_\mu \frac{U_{Sg}}{R_S}$$

In accordance with the diagram shown in figure 3 the transfer function for the DFM speed in the coordinate system associated with the stator representing vector at balancing emf and the constant voltage on the stator has the form:

$$\frac{\omega(p)}{u_{Ri}(p)} = \frac{3U_{Sg} z_P L_\mu}{2JR_R R_S T_R p^2 + 2JR_R R_S p + 3\frac{\omega_g}{R_S \omega_0} U_{Sg}^2 z_P L_\mu^2} =$$

$$= \frac{\frac{1}{\left(\frac{U_{Sg} L_\mu \omega_g}{R_S \omega_0}\right)}}{\frac{2JR_R R_S T_R}{3\frac{\omega_g}{R_S \omega_0} U_{Sg}^2 z_P L_\mu^2} p^2 + \frac{2JR_R R_S}{3\frac{\omega_g}{R_S \omega_0} U_{Sg}^2 z_P L_\mu^2} p + 1}.$$

By introducing the DFM mechanical time constant $T_M = \frac{2JR_R R_S}{3\frac{\omega_g}{R_S \omega_0} U_{Sg}^2 z_P L_\mu^2}$, and the equivalent DFM transfer factor $k_E = \dfrac{1}{\left(\dfrac{U_{Sg} L_\mu \omega_g}{R_S \omega_0}\right)}$, we get

$$\frac{\omega(p)}{u_{Ri}(p)} = \frac{k_e}{T_M T_R p^2 + T_M p + 1}. \qquad (5)$$

The transfer function of the moment with respect to the rotor voltage, taking into account the above notations, has the form:

$$\frac{M(p)}{u_{Ri}(p)} = \frac{{}^{T_M}\!/_J \left(T_R p + 1\right)}{T_M T_R p^2 + T_M p + 1}. \qquad (6)$$

It follows from expressions (5) and (6) that, with the assumptions made, the transfer functions for the velocity and electromagnetic moment of the machine have a simple form convenient for use in the electric drive control systems.

3. Conclusion

Thus, the application of the rotating coordinate system associated with one of the machine coordinates, as well as the balancing of the mutual influence of the windings emf of the motor rotor and stator like phases, make it possible to simplify considerably the mathematical model of the dually-fed machine and bring it closer to the mathematical model of the DC motor with an independent excitement. This allows the operation of the machine to be analyzed and its active rotor current and electromagnetic moment to be easily identified, which is important for practical purposes.

References

[1] Onischenko G B and Lokteva I L 1979 *Asynchronous Valve Cascades and Dually fed Motors* (Moscow: Energia) p 200

[2] Ostrovlyanchik V Yu and Popolzin I Yu 2016 *Proc. of Int. Sci. Conf. in Knowledge-based Technologies of Development and Use of Mineral Resources* (Novokuznetsk: SibSIU) vol 2 pp 303–309

IOP Conf. Series: Materials Science and Engineering **354** (2018) 012018 doi:10.1088/1757-899X/354/1/012018

Development of the pump protection system against cavitation on the basis of the stator current signature analysis of drive electric motor

M V Kipervasser, A V Gerasimuk and V P Simakov

Siberian State Industrial University, 42 Kirova street, Novokuznetsk, 654007, Russia

E-mail: kipervasser2012@yandex.ru

Abstract. In the present paper a new registration method of such inadmissible phenomenon as cavitation in the operating mode of centrifugal pump is offered. Influence of cavitation and extent of its development on the value of mechanical power consumed by the pump from the electric motor is studied. On the basis of design formulas the joint mathematical model of centrifugal pumping unit with the synchronous motor is created. In the model the phenomena accompanying the work of a pumping installation in the cavitation mode are considered. Mathematical modeling of the pump operation in the considered emergency operation is carried out. The chart of stator current of the electric motor, depending on the degree of cavitation development of is received. On the basis of the analysis of the obtained data the conclusion on the possibility of registration of cavitation by the current of drive electric motor is made and the functional diagram of the developed protection system is offered, its operation principle is described.

1. Introduction

Centrifugal pumping installations are the most often used units at the enterprises of different technology orientation [1]. Such pumping installations are widely used in different brunches of industry and life support: in water disposal systems of mines, open-pit mines, pits, in urban pumping stations (also cold water supply, the sewerage), at combined heat and power plant, state district power plant, the NPP (circulating, feeding, condensate, etc. pumps), at the plants of petrochemical, metallurgical, food industry for pumping of different liquids.

In some cases centrifugal pumps are especially responsible units. These include pumps for urban water supply systems, circulating pumps for nuclear power plants, drainage of mining enterprises, pumps for cooling systems for blast furnaces. High requirements to reliability and uninterrupted operation of work cause need of development and designing of protection systems of pumping installations against abnormal and emergency operation of work.

It is reasonable that the system of protection includes the minimum quantity of elements, in particular – sensors. In the existing schemes of control and protection sensors are the most vulnerable elements of the system from the point of view of failure.

The requirement to minimize the number of elements can be met by a system created on the basis of the method of indirect diagnostics of the equipment operation – stator current control of the drive motor. Such a system in its composition will have only the protection device itself (based on a microprocessor terminal or some other electronic device) and a measuring transducer.

Content from this work may be used under the terms of the Creative Commons Attribution 3.0 licence. Any further distribution of this work must maintain attribution to the author(s) and the title of the work, journal citation and DOI.
Published under licence by IOP Publishing Ltd

One of the most dangerous for centrifugal pumping units is the mode of cavitation. Cavitation is a violation of the continuity of the fluid that occurs in sections of the stream where the pressure, falling, reaches a certain critical value [2]. When the pump sucks liquid from the reservoir, the pressure in the suction pipeline of the P_{IN} drops below atmospheric pressure [3].

In figure 1 emergence of cavitation in the pump is shown. In the pipeline the pressure gradient takes place. At the beginning of the pipeline pressure is close to atmospheric, and the closer to the pump – the discharge is greater. In process of advance of liquid in the pump its pressure can become less than the pressure of the liquid saturated steam under existing conditions.

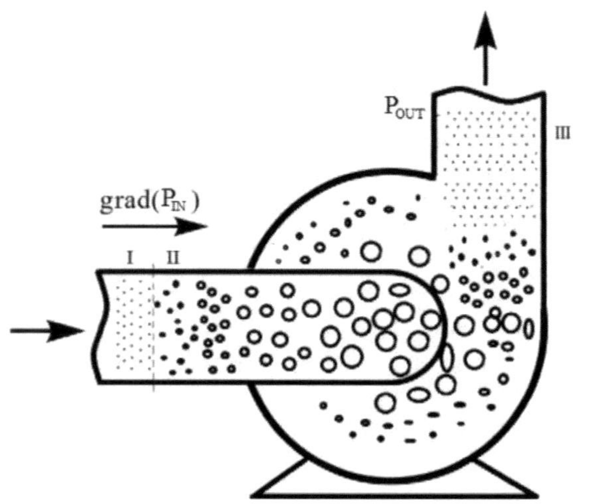

I - a zone of reduced pressure without cavitation; II - a zone of reduced pressure with cavitation; III - a pressure zone

Figure 1. Process of liquid cavitation in the pump.

In this case there is a cold boiling up of liquid. Along with it there is an emission of the dissolved gas from the liquid. The bubbles filled with saturated steam are formed. The lower the pressure drops, the more bubbles increase in volume, turning into large bubble-caverns. Further, the bubbles are carried away into the zone of high pressure (forcing pressure) of P_{OUT} beginning at the exit from a pump wheel where instantly disappear due to condensation of the vapors filling them. Thus, in the flow there is a fairly narrowly limited cavitation zone. This phenomenon is characterized by vibrations and noises. With a strong development of cavitation, flow disruption and pump shutdown can occur [1, 2].

When the caverns are collapsed on the surface of the impeller and the pump volute, a force attack – a hydroblow occurs. In this case, a pressure develops, the value of which has values exceeding the yield strength of the wheel and pump volute materials. The prolonged impact of cavitation leads to the destruction of the pump.

2. Modeling of behavior of stator current of the electric motor at cavitation

It is known that resistance moment on the shaft of the engine which is created by the pump can be determined by the following formula [2, 4]:

$$M_C = \frac{30}{\pi n} \frac{Q \rho g H}{\eta}, \quad (1)$$

where Q – pump flow, m^3/h; ρ – liquid density, kg/m^3; H – pump pressure, m; η - efficiency of the pump, n - rotating speed of the pump, RPM.

Also, it is known that when the positive suction head of the pump decreases, its head drops and its efficiency decreases, which is confirmed by the kink of the experimental curve given in [1, 3] (zone II).

As it can be seen from the graph, as the positive suction head decreases, the power consumed by the pump first increases and then falls sharply to the values close to the ones of the dry run, since with fully developed cavitation the flow breaks down and the pump stops pumping the liquid [1, 3].

On the basis of the above it is possible to draw a conclusion that current of the drive electric motor of the pump will change according to the change in the loading on its shaft [5, 6].

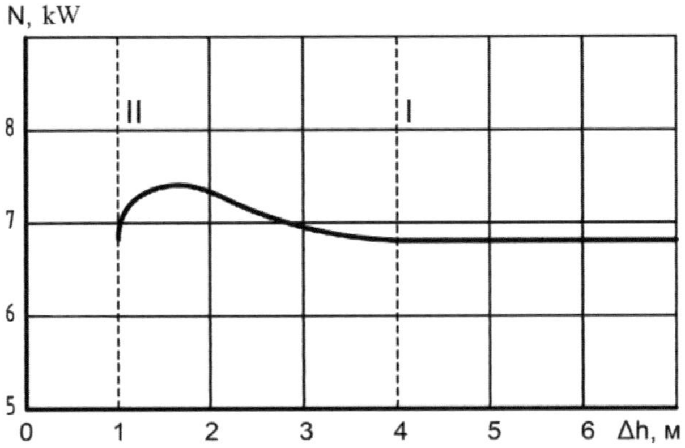

Figure 2. Dependence of the mechanical power consumed by the pump depending on a positive suction head.

For confirmation of this assumption in the software MatLAB Simulink modeling of work of pumping installation in the cavitation mode was carried out.

As a model of pumping installation the unit as a part of the pump AD4000-95-2S was used (pump flow – 4000 m^3/h, pressure – 95 m) and synchronous motor SND 2-16-59-6U3 (1600 kW, 1000 RPM, 6 kV). The model of the engine is implemented on a two-phase system of the Park-Gorev' equations [7].

$$\begin{cases} i_a = i_d \cos\gamma - i_q \sin\gamma \\ i_b = i_d \cos\left(\gamma - \dfrac{2\pi}{3}\right) - i_q \sin\left(\gamma - \dfrac{2\pi}{3}\right) \\ i_c = i_d \cos\left(\gamma + \dfrac{2\pi}{3}\right) - i_q \sin\left(\gamma + \dfrac{2\pi}{3}\right) \\ U_d = \dfrac{d\psi_d}{dt} - \omega\psi_q + r_1 i_d \\ U_q = \dfrac{d\psi_q}{dt} - \omega\psi_d + r_1 i_q \\ U_f = \dfrac{d\psi_f}{dt} + r_f i_f \\ 0 = \dfrac{d\psi_{yd}}{dt} + r_{yd} i_{yd} \\ 0 = \dfrac{d\psi_{yq}}{dt} + r_{yq} i_{yq} \\ M = M_C + J\dfrac{d\omega}{dt} \\ M = \psi_d i_q - \psi_q i_d \\ \omega_1 - \omega = \dfrac{d\theta}{dt} \end{cases} \qquad (2)$$

where i_a, i_b, i_c – phase currents of the stator, A; U_d, U_q – voltage on stator winding clips on longitudinal and cross axes; Ψ_d, Ψ_q, i_d, i_q – according to flux linkage and currents of windings of phases of the stator on longitudinal and cross axes; r_1 – the active resistance of phase winding of the stator; U_f – voltage on field winding clips; Ψ_f, i_f – flux linkage and current of field winding; r_f – the active resistance of field winding; Ψ_{yd}, Ψ_{yq}, i_{yd}, i_{yq} – according to flux linkage and currents of damper winding on longitudinal and cross axes respectively; r_{yd}, r_{yq} – active resistance of damper winding on longitudinal and cross axes; M – the electromagnetic moment, M_s – the moment of loading resistance, J – inertia moment of rotating masses, γ – the angular position of a rotor.

The system of the equations (4) accurately describes operation of the synchronous motor at a symmetric mode, is applicable to the considered case and is implemented in the form of the standard block from Simulink/Simscape/SimPowerSystems library.

The functional diagram of the model of pumping unit in the mode of cavitation is provided on the figure 3.

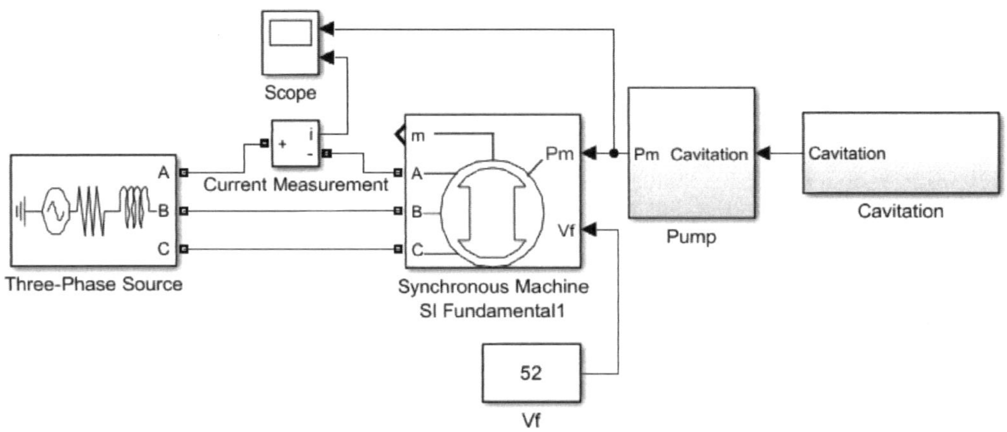

Figure 3. The functional diagram of the model of pumping unit in the mode of cavitation.

3. Results of modeling

The results of modeling of pump unit performance during the occurrence of the phenomenon of cavitation are given in the figure 4.

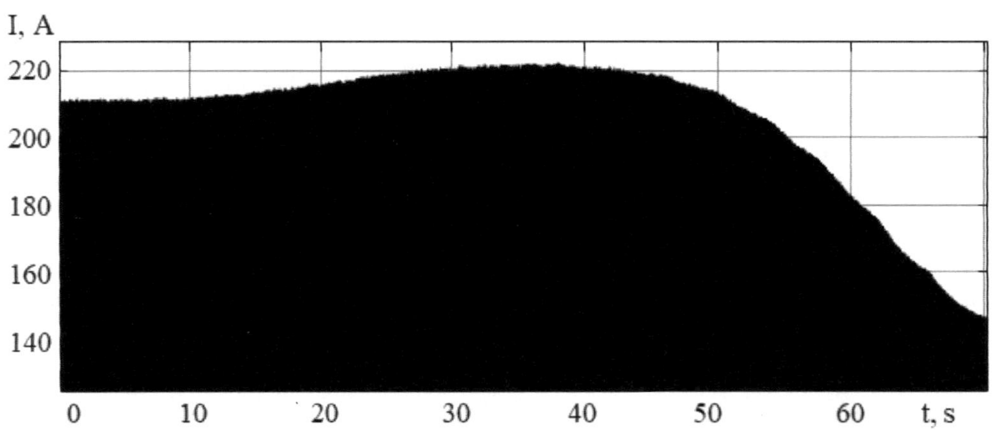

Figure 4. Diagram of the stator phase current of pump drive engine working during cavitation.

The diagram of the stator current, as expected, repeats with its own form the load curve on the motor shaft during cavitation.

Cavitation, which has just begun to develop, is technically difficult to detect with respect to the stator current, since some increase in the stator current in the initial section of zone II of figure 3 will be virtually indistinguishable from the increase in current caused by the increase of the torque on the shaft due to technological overload. However, the most dangerous developed cavitation can be identified by this method. Rapidly developing cavitation causes a drop in the load power on the motor shaft, and hence the phase current of the stator. In this case, cavitation can be identified by the rate of change of the current value of the stator current. The faster the cavitation develops, the faster the stator current falls. Cavitation, which develops slowly, can be detected by comparing the stator current in the normal mode of the pump with current in cavitation mode.

4. Development of the functional diagram of system of protection

The offered functional diagram of the protection system of pumping equipment against cavitation is presented in figure 5.

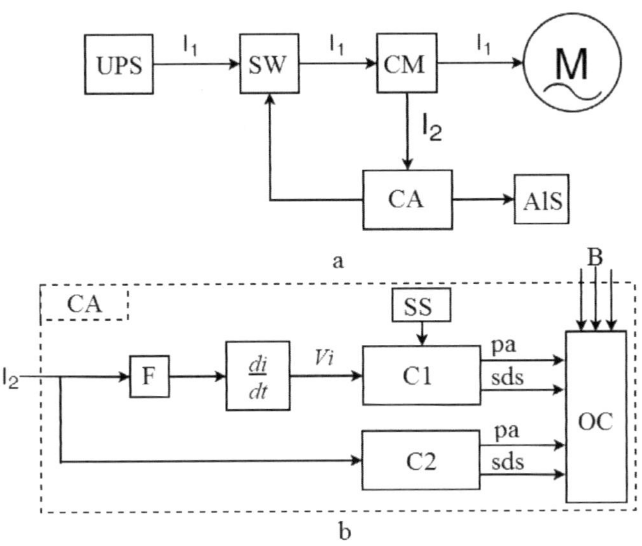

UPS – a utility power system, SW – the switch, CM – the current measuring instrument, CA – the device of the current analysis, AlS – the alarm system, M – the electric motor, F – the filter, SS – a set of settings, C1, C2 – comparators, VTs – output circuit, I1 – primary current of the stator, I2 – secondary current of the stator, Vi – a signal of change rate of the current virtual value, sds – a shutdown signal, pa – a preventive alarm, B – signals of technology blocking.

Figure 5. Functional diagram of the protection system of the pump from cavitation on stator current (a) and device of the current analysis (b).

The principle of system operation is the following: the measuring instrument of current will transform primary strengths of stator current to a virtual value and gives it to the device of the current analysis.

The analysis device has in its composition two comparison channels. One channel compares the current stator current with the current value in the normal mode (set in the setting block when adjusting the equipment). If the difference between the specified stator current and the measured current is not large, a warning signal is given to the duty personnel on the abnormal operating mode of the pumping unit. If the difference is significant, then the second stage of protection is triggered – shutdown of the pump, if this is not prohibited by technological interlocks. The second channel calculates the derivative of the current value of the stator current in time, that is, determines the rate of change in the stator current. The filter cuts off the high-frequency part of the current oscillations caused by noise, jolts of the current due to any processes in the mains or any other phenomena, since they can cause a false protection trigger. The obtained rate of current change is compared with a set of preset settings that determine the further operation of the device. If the rate of change is small, a warning signal is given to the duty personnel on the abnormal operating mode. If the rate of change is high and the stator current falls rapidly, a signal is sent to turn off the pump.

IOP Conf. Series: Materials Science and Engineering **354** (2018) 012018 doi:10.1088/1757-899X/354/1/012018

5. Conclusion

The offered system, thanks to a small number of structural elements will exceed other existing protection systems in terms of reliability. Besides, owing to the use of the stator current of the drive engine as a control parameter, the high-speed performance will also be greater, than at the existing systems.

References

[1] Karelin V Ya and Minayev A B 1986 *Pumps and Pumping Stations* (Moscow: Stroyizdat) p 320
[2] Abdurashitov S A and Tupichenkov A A 1974 *Pumps and Compressors* (Moscow: Subsoil) p 296
[3] Turk V I and Minayev A B 1976 *Pumps and Pumping Stations* (Moscow: Stroyizdat) p 304
[4] Shlipchenko Z C 1976 *Pumps, Compressors, Fans* (Kiev: Tekhnika) p 386
[5] Pugachyov E V, Kipervasser M I and Gerasimuk A V 2016 *Bulletin of KuzSTU* **2** 84–90
[6] Pugachyov E V, Kipervasser M I and Inzhelevsky O V 2011 *Bulletin of KuzSTU* **4** 51–55
[7] Sipaylov A 1987 *Electrical Machines (Special Course)* (Moscow: The Higher School) p 287

Features of the use of time-frequency distributions for controlling the mixture-producing aggregate

D B Fedosenkov[1], A A Simikova[2] and B A Fedosenkov[2]

[1]Siberian Generating Company, 53 build. 5, Doubininskaya Street, Moscow, 115054, Russia
[2]Kemerovo State University, 6 Krasnaya Street, Kemerovo, 650000, Russia

E-mail: rafwavelet@ngs.ru

Abstract. The paper submits and argues the information on filtering properties of the mixing unit as a part of the mixture-producing aggregate. Relevant theoretical data concerning a channel transfer function of the mixing unit and multidimensional material flow signals are adduced here. Note that ordinary one-dimensional material flow signals are defined in terms of time-frequency distributions of Cohen's class representations operating with Gabor wavelet functions. Two time-frequencies signal representations are written about in the paper to show how one can solve controlling problems as applied to mixture-producing systems: they are the so-called Rihaczek and Wigner-Ville distributions. In particular, the latter illustrates low-pass filtering properties that are practically available in any of low-pass elements of a physical system.

1. Introduction

The mixture-producing aggregate that is the object of interest, consists of different fragments: continuous- and discrete-type feeding devices, transferring and material mass-forming equipment, a mixing system proper with multi-channel feedbacks, and other engineering appliances. To control such a mixture-producing structure, it is necessary both to realize the function of controlling and monitoring all of the material flows in every technological point of the unit. This is dealt with the computer-aided control system based on the digital wavelet transforms and time-frequency representations of Cohen's class. Thus, the entire digital control system with the mixture-producing unit as a technological plant makes up the controlled mixture-producing aggregate. The objective of the paper is to show the presence of filtering effects in the aggregate, and we begin with the so-called channel Cohen's class transfer function of the mixing unit formed of input and output material flow signal multidimensional representations, *Coh*-functions of lapsed time and instantaneous frequency.

2. Mixing unit: time-frequency material flow signals

To obtain the output (with respect to the mixing system of the mixture-producing aggregate) modified material flow signal having the form of a time-frequency distribution (one of Cohen's class distributions [1]), we introduce and denote the channel transfer function, time-variant in its structure, as

$$H(t,\omega)=Coh\{x_O(t)\}/Coh\{x_I(t)\},$$

Content from this work may be used under the terms of the Creative Commons Attribution 3.0 licence. Any further distribution of this work must maintain attribution to the author(s) and the title of the work, journal citation and DOI.
Published under licence by IOP Publishing Ltd

IOP Conf. Series: Materials Science and Engineering **354** (2018) 012019 doi:10.1088/1757-899X/354/1/012019

where $Coh\{x_i(t)\}$ and $Coh\{x_o(t)\}$ are respectively the time-frequency distributions of Cohen's class for the input and output one-dimensional material flow signals. Here multidimensional distributions of Cohen's class [1, 2] have been obtained on the basis of Gabor wavelet functions [3, 4]. Then the distribution of the mixing device output signal is written as

$$Coh_0(t,\omega) = H(t,\omega)Coh_I(t,\omega),$$

where $Coh_I(t,\omega) \equiv Coh\{x_I(t)\}$; $Coh_O(t,\omega) \equiv Coh\{x_O(t)\}$.

The output one-dimensional signal can be defined as the following convolution integral:

$$x_O(t) = \int_0^t S_I(t')h(t,t')dt',$$

where $h(t,t')$ is the shifted-in-time impulse response characteristic (the weight function) of the mixing device, in other words, $h(t,t')$ is "the window" for the input signal, t is the current time, t' is the lapsed running time (i.e., the time that is being shifted), $S_I(t')$ is the input signal with being-shifted time instants taken into account. In this case, the connection between time-frequency modified input and output signals can be written as follows:

$$Coh_O(t,\omega) = \int_0^t \int_0^\omega K(t,\omega;t',\omega')Coh_I(t',\omega')dt'd\omega',$$

where $Coh_I(t',\omega')$ is the input multidimensional (MD) time-frequency signal, ω' is the frequency corresponding to the being-shifted time, and $K(t, \omega, t', \omega')$ is the window function of an output MD-distribution (one or other).

Direct calculation gives the expression

$$K(t,\omega;t',\omega') = \frac{1}{8\pi^3}\int_o^t \exp\{-j[\omega\tau - \omega'\tau' - \theta(u-t) + \theta'(u'-t')]\}\frac{\Phi(\theta,\tau)}{\Phi(\theta',\tau')}\times$$
$$\times h^*\left(u-\frac{\tau}{2},u'-\frac{\tau'}{2}\right)h\left(u+\frac{\tau}{2},u'+\frac{\tau'}{2}\right)du d\tau d\theta du' d\tau' d\theta'$$

Here, $\Phi(\theta, \tau)$ is a time-frequency distribution kernel that can change a distribution proper.

For example, for the Rihaczek distribution [1], we have:

$$K(t,\omega;t',\omega') = \frac{1}{2\pi}h(t,t')S^*(\omega,\omega')\exp(-j\omega t + j\omega' t'),$$

where $S(\omega,\omega') = \frac{1}{2\pi}\int_0^\tau\int_0^{\tau'} h(t,t')\exp(-j\omega t - j\omega' t')dtdt'$ is the spectral function of the mixer impulse response. Here $K(t, \omega, t', \omega')$ is a two-dimensional Rihaczek distribution.

For the Wigner distribution we obtain:

$$K(t,\omega;t',\omega')=\frac{1}{2\pi}\int_0^t\int_0^{t'}\exp(-j\omega\tau+j\omega'\tau')h^*\left(t-\frac{\tau}{2},t'-\frac{\tau'}{2}\right)h\left(t+\frac{\tau}{2},t'+\frac{\tau'}{2}\right)d\tau d\tau',$$

which is a two-dimensional Wigner distribution [1].

3. Filtering features of the mixing unit

In order to control the parameter of smoothing the material flow fluctuations at the pre-mixing stage, that is extremely important from the point of view of possibility to control the process of obtaining high-quality mixtures, time-frequency maps (TF-maps) at the input (figure 1) and the output (figure 2) of the mixing device were registered.

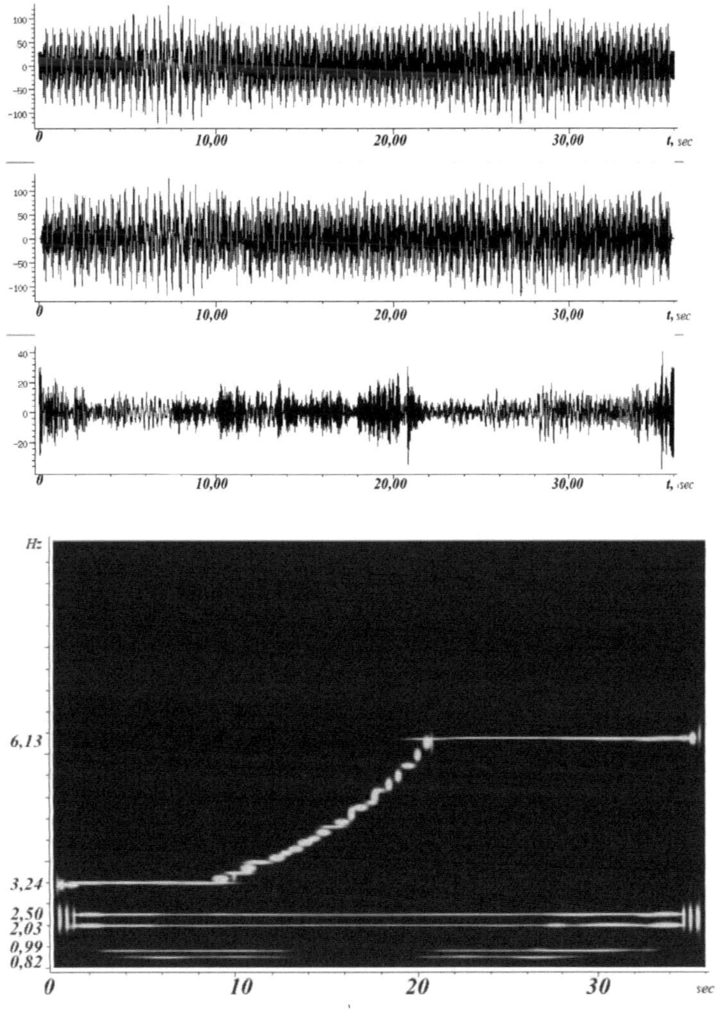

Figure 1. The actual flow signal; the signal reconstructed according to the wavelet-matching pursuit algorithm [4]; the error of adaptive approximation, and the signal TF-map at the pre-mixing stage.

Under those conditions, the feeding devices unit output signal was being formed by feeders of the following types:

1) two screw feeders operating in the portion-type feeding mode: frequency operation modes are of 0.82 Hz (the dosing period T = 1.22 sec) and 0.996 Hz (T = 1.13 sec);
2) two spiral feeders operating in the beat oscillation mode: operating frequencies are such ones - 2.03 Hz (T = 0.49 sec) and 2.50 Hz (T = 0.39 sec); in this case, the total waveform of these two signals contains a resulting low-frequency component of a frequency equal to the difference in the operating frequencies of the feeders. As it can be seen from figure 1 and figure 2, these two signals are clearly separated on the time-frequency map;
3) a spiral feeder when one readjusting its working regime (a certain load drop mode) and, specifically, changing the feeder operating mode from 3.24 Hz (T = 0.30 sec) up to 6.13 Hz (T = 0.16 sec).

Effective control of the mixing process dynamics is possible on the basis of a complex analysis of material flow signals in the temporal and time-frequency domains. For these purposes, the material flow signals recorded at the pre-mixing stage, the mixing stage (signals in the global and local recycle-channels), and at the outlet of the mixture-producing aggregate (MPA) are suitable.

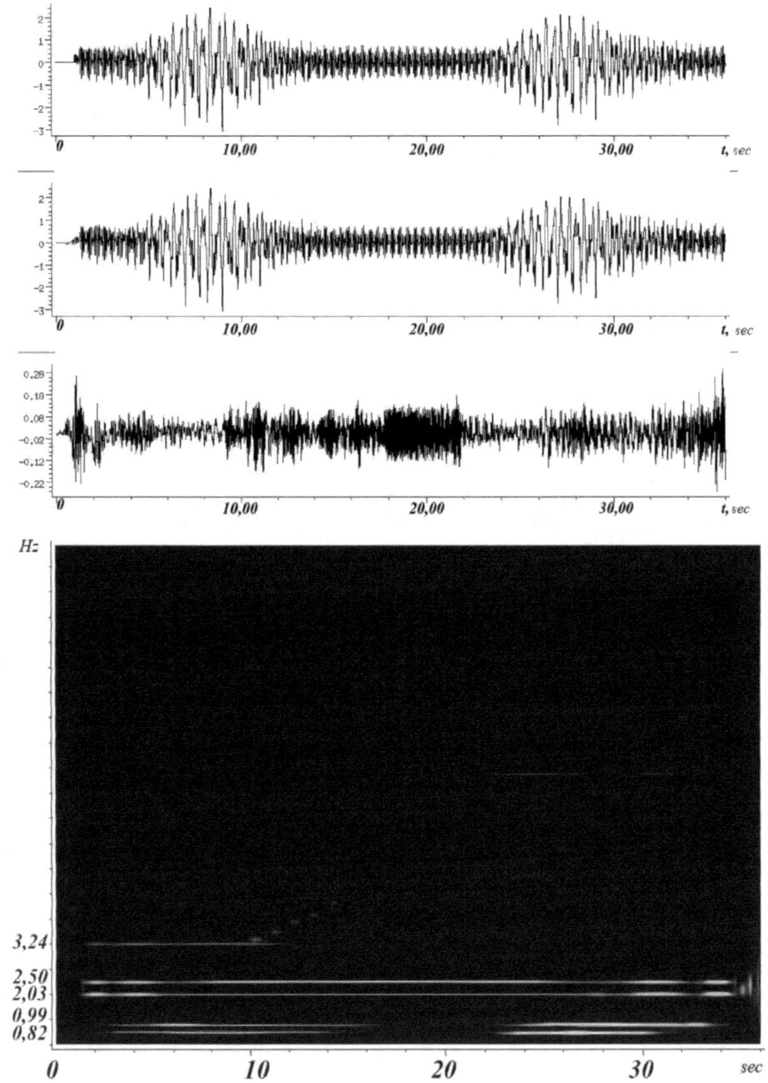

Figure 2. The same (see figure 1) at the mixing device output.

Since the mixing equipment (mixing device and purposefully organized flow channels) are approximated for simulation by the first / second-order links and have the properties of a low-pass filter, then certain high-frequency components of a material flow signal, occurring at the feeding stage, may be absent in the signal spectrum at the output of the MPA and / or in the local and global recycle-channels. Consequently, they can not be used for control purposes, i.e., the material flow signals recorded at the output of mixing devices are pertinent to control feeding devices running at relatively low frequencies available in the spectrum of a material flow signal. Thus, e.g., the analysis of figure 2 shows that at the output of the mixing system there are no time-frequency atoms corresponding to the mode of readjusting the working performance of the spiral feeding device (see figure 1). Only the time-frequency atoms at frequencies not exceeding 3.3 Hz, are distinctly seen to be identified on the Wigner map.

Thus, from the point of view of the effectiveness to control the dynamics of the mixture-producing process, the optimal decision is to use modified variables in the form of time-frequency maps at the pre-mixing and post-mixing stages.

4. Conclusion

It is shown on the example of a mixture-producing aggregate, that time-frequency distributions of Cohen's class are transformed by the devices of technological equipment possessing low-frequency properties, with the effect of low-frequency filtration - like one-dimensional material flow signals. And in that case informative clarity of one-dimensional material flow signals at the post-mixing stage (see figure 2), i.e. at the mixing device output, does not go to any comparison with that rich semantics which corresponds to the multidimensional time-frequency representation (here, the Wigner map). On such a representation low-frequency elements of the post-mixing map are fully saved in an initial kind corresponding to their form and localization on the pre-mixing map (see figure 1). At the same time, elements (time-frequency atoms) on the mixer output Wigner map defining the process of readjusting a working regime (a certain load drop mode) of one of feeding devices from the operating frequency of 3.24 Hz to higher frequencies, distinctly show up merely in the frequency range where the mixer gain-frequency characteristic has relatively high values. A similar conclusion can be spread on other technological devices that, as to their properties, look like low-frequency filters, i.e. devices which dynamics is described by the models of relevant inertial links.

References

[1] Cohen L 1995 *Time-Frequency Analysis* (Englewood Cliffs: Prentice Hall) p 299
[2] Choi H I and Williams W J 1989 *IEEE Trans. Acoust., Speech, and Signal Proc.* **37(6)** 862–71
[3] Daubechies I 1992 *Ten Lectures on Wavelets: Regional Conference Series in Applied*
[4] *Mathematics* (Philadelphia PA: CBMS-NSF SIAM) p 357
[5] Fedosenkov B A, Nazimov A S and Sheboukov A V 2004 *Automation and Modern Technologies: Automation of Scientific Research and Production Processes* **8** 7–13

Institute of Physics
Dirac House, Temple Back
Bristol BS1 6BE UK

ISSN: 1757-8981
ISBN 978-1-5108-6284-5

9 781510 862845

CURRAN ASSOCIATES INC.
proceedings
.com